BASIC GEOLOGY OF THE SANTA MARGARITA AREA, SAN LUIS OBISPO COUNTY, CALIFORNIA

by Earl W. Hart

BULLETIN 199

1976

CALIFORNIA DIVISION OF MINES AND GEOLOGY
1416 9TH STREET, ROOM 1341
SACRAMENTO, CA 95814

THE GREAT SEAL OF THE STATE OF CALIFORNIA
EUREKA
CALIFORNIA

STATE OF CALIFORNIA
EDMUND G. BROWN JR., *GOVERNOR*

THE RESOURCES AGENCY
CLAIRE T. DEDRICK, *SECRETARY FOR RESOURCES*

DEPARTMENT OF CONSERVATION
LEWIS A. MORAN, *DIRECTOR*

DIVISION OF MINES AND GEOLOGY
THOMAS E. GAY JR., *ACTING STATE GEOLOGIST*

CONTENTS

ABSTRACT 5
INTRODUCTION 7
ROCK UNITS 8
    Pre-Tertiary rocks west of Rinconada fault 8
        Franciscan melange 8
        Ultramafic-mafic complex 11
        Mafic volcanic rocks 12
        Toro Formation 13
        Atascadero Formation 14
    Pre-Tertiary rocks east of Rinconada fault 16
        Metamorphic rocks 16
        Granitic rocks 16
        Unnamed sandstone and conglomerate 18
    Cenozoic rocks 19
        Unnamed conglomerate 19
        Simmler Formation(?) 20
        Vaqueros Formation 20
            West of the Rinconada fault 21
            East of the Rinconada fault 21
                Unnamed conglomerate member 22
                Thickness and conditions of deposition 22
                Age and correlation 22
        Monterey Formation 26
            Central synclinal belt 27
                Lower member 27
                Upper member 28
            East of the Rinconada fault 28
            Western area 29
            Origin and conditions of deposition 29
            Age and correlation 29
        Santa Margarita Formation 31
        Paso Robles Formation 32
        Older alluvium 33
        Younger alluvium 34
        Landslide deposits 34
STRUCTURAL FEATURES 34
    Faults 35
        Rinconada fault 35
        "Nacimiento" fault zone 36
            Southwest-dipping faults 36
            Vertical faults 36
            Other fault elements of the "Nacimiento" zone 37
        Other faults 37
            Faults west of the "Nacimiento" fault zone 37
            Faults between the "Nacimiento" and Rinconada faults 37
            Faults east of the Rinconada fault 38
            Middle Branch fault 38
    Folds 38
        Santa Margarita syncline 38
        Other folds 38
SUMMARY OF GEOLOGIC HISTORY 39
    Late Mesozoic (Jurassic-Cretaceous) time 39
    Latest Cretaceous to mid-Tertiary time 39
    Oligocene(?) time 39
    Late Cenozoic time 39

MINERAL RESOURCES 39
    Carbonate rocks 41
    Clay 42
    Manganese 42
    Mineral springs 42
    Petroleum 42
    Sand and gravel 43
    Specialty sand 43
    Stone 43
REFERENCES 44

ILLUSTRATIONS

Plate 1. Geologic map and sections, Santa Margarita area .... (pocket)

Figure 1. Location map 7
Figure 2. Chart of time-stratigraphic (time) units 8
Figure 3. Composite stratigraphic column west of Rinconada fault 9
Figure 4. Composite stratigraphic column east of Rinconada fault 10
Figure 5. Ternary diagram—composition of granitic rocks 17
Figure 6. Correlation chart of middle Tertiary rock units 26

Table 1. Megafossils of the Toro Formation 14
Table 2. Fossils of the Atascadero Formation 16
Table 3. Potassium-argon age determinations of biotite in granitic rocks 17
Table 4. Upper Cretaceous microfossils east of Rinconada fault 19
Table 5. Check list of fossil foraminifers, Monterey and Vaqueros Formations 23
Table 6. Megafossils of the Monterey and Santa Margarita Formations 30
Table 7. Upper Miocene diatoms 31
Table 8. Mineral deposits and prospects 40
Table 9. Exploratory wells drilled for petroleum 42

ABSTRACT

The Santa Margarita area of investigation lies astride the Rinconada fault, which divides the area into two distinct geologic and physiographic terranes—the La Panza Range to the east and the Santa Lucia Range to the west. The La Panza Range consists of Cretaceous granodiorite and adamellite with minor associated schist of the Salinian block. This is overlain by unnamed Upper Cretaceous sandstone and conglomerate and a relatively thin sequence of marine and nonmarine sedimentary rocks of Oligocene(?) to Holocene age. In contrast, "basement" rocks of the Santa Lucia Range consist of the chaotic Franciscan melange—a tectonic unit of probable late Mesozoic age. Overlying this are thrust plates of: 1) serpentinite and an intrusive complex of ultramafic-mafic rocks, 2) mafic volcanic rocks with chert, and 3) incomplete Great Valley-like sedimentary sequences of the Toro and Atascadero Formations of Late Jurassic to Late Cretaceous age. Tertiary marine and nonmarine sequences of sedimentary and volcanic rocks rest unconformably on the older units—the best preserved sequence being the Vaqueros, Monterey, and Santa Margarita Formations of Miocene age along the northeast margin of the Santa Lucia Range.

Aside from being stratigraphically complex, most of the formations are severely faulted and folded as a result of two or more prolonged periods of tectonism. Late Mesozoic to early Tertiary thrust faulting is largely obscured by northwest-trending faults and folds of Miocene to Pleistocene (Holocene?) age. The principal structures are the Rinconada and "Nacimiento" fault zones and the intervening Santa Margarita syncline.

The importance of the Rinconada fault is manifest in the highly contrasting rock units that are truncated by and juxtaposed along the fault. Correlation of a unique giant-boulder conglomerate (Simmler Formation?) and associated marine beds of the Vaqueros Formation with a similar apparently offset sequence of rocks 33 miles to the northwest suggests large-scale right-lateral displacement along the Rinconada fault and related connecting faults. Slivers of Franciscan rocks exposed along the southern segment of the Rinconada fault suggest that the fault is the local boundary between the Salinian and Franciscan (Nacimiento) blocks and that the "Nacimiento" fault zone lies totally within the Franciscan terrane in the map area.

The ill-defined "Nacimiento" fault zone extends the length of the map area (18 miles) and appears to truncate the Rinconada fault at an acute angle at the east margin of the area. The "Nacimiento" zone consists of two principal fault elements—thrust and reverse faults that commonly dip southwest and near vertical faults. Although not unique to the map area, the southwest-dipping faults are in sharp contrast to the principal northeast-dipping faults reported to the northwest and southeast along the so-called Sur-Nacimiento fault zone (Vedder and Brown, 1968; Page, 1970a).

At least 24 mineral deposits have been prospected in the map area, but only crushed and broken stone and sand and gravel have been produced in significant amounts. The Santa Margarita granite quarry is the principal source of high-quality aggregate and riprap in San Luis Obispo County.

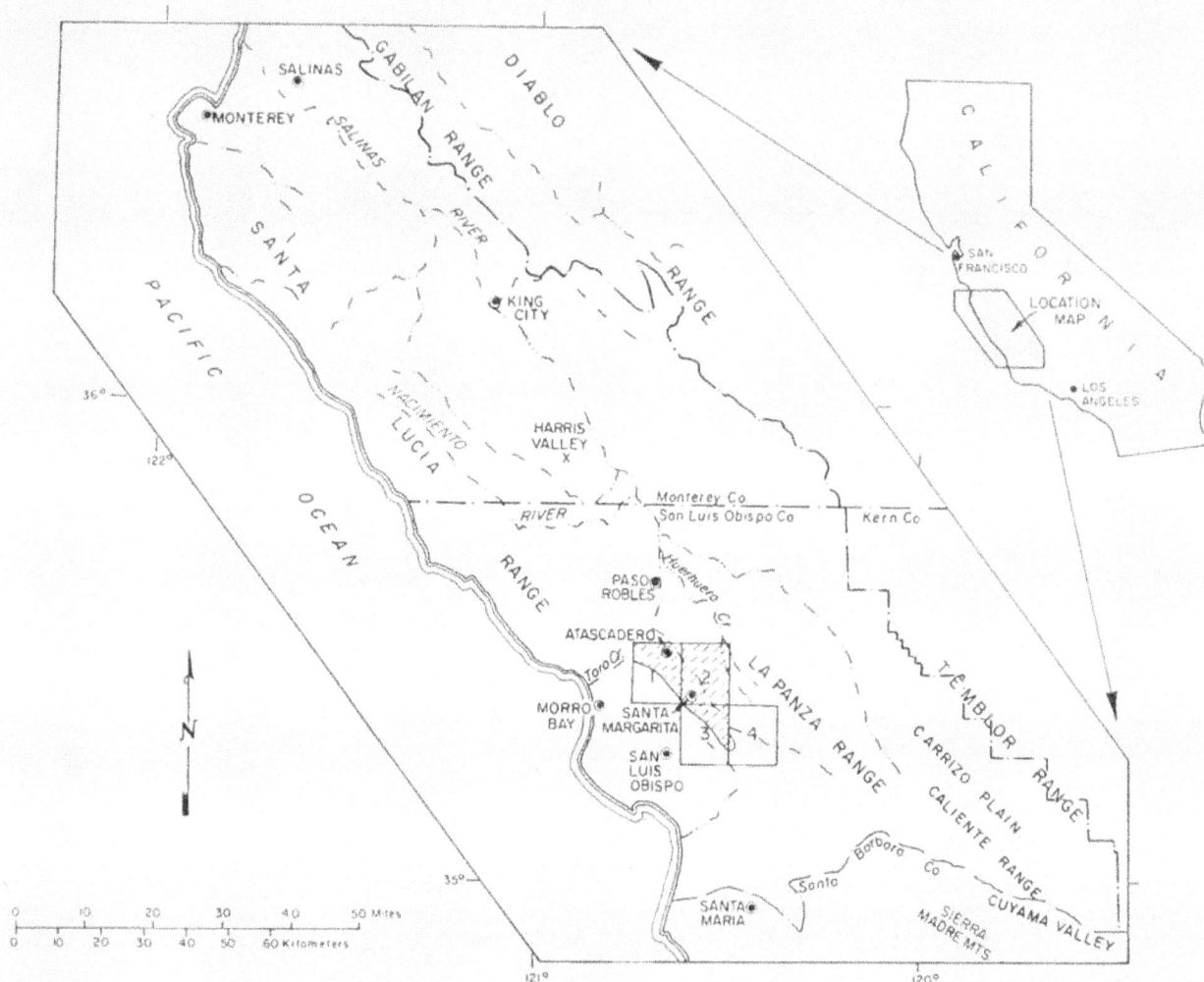

Figure 1. Location map of the Santa Margarita study area showing principal geographic features and 7.5-minute quadrangles mapped—(1) Atascadero, (2) Santa Margarita, (3) Lopez Mountain, (4) Pozo Summit.

BASIC GEOLOGY OF THE SANTA MARGARITA AREA, SAN LUIS OBISPO COUNTY, CALIFORNIA

by Earl W. Hart[1]

INTRODUCTION

The Santa Margarita area lies in the southern Coast Ranges midway between Los Angeles and San Francisco. The area of investigation covers 126 square miles in parts of the Santa Lucia and La Panza Ranges and the intervening valley and rolling hill area of the Salinas River drainage (figure 1). Atascadero is the main population center and, although unincorporated, has a population of more than 16,000. Agriculture, including cattle raising, is the principal industry. Access is good, except for the interior parts of the Santa Lucia and La Panza Ranges, which are rugged and densely vegetated.

Field work was conducted intermittently between fall 1965 and spring 1971. Mapping was done at a scale of 1:24,000 on the Atascadero, Santa Margarita, Lopez Mountain, and Pozo Summit 7 1/2-minute topographic quadrangles. More than 300 samples were collected for fossil identification, thin-section examination, feldspar staining (sandstones, granites), and x-ray diffraction and fluorescence analyses. Much of these data are included herein.

This investigation was based on the need for detailed geologic data on rock units and faults of the southern Coast Ranges, little of which was available in 1965. Little was known of the upper Mesozoic rocks of this region and data on Tertiary stratigraphy and lithology were rather general.

[1]Geologist, California Division of Mines and Geology

Information was also needed on the nature and ages of the "Nacimiento" and Rinconada faults, neither of which had ever been mapped in detail. Moreover, the Salinian and Franciscan basement rocks are nearly juxtaposed, and the mapped area is the only locality where nearly adjacent exposures of these two contrasting basement terranes can be studied southeast of the Sur-Nacimiento fault zone exposures in southwestern Monterey County.

The only comprehensive mapping of the Santa Margarita area was done by Fairbanks (1904) at a scale of 1:125,000. Since then, a small area near the junction of the Rinconada and "Nacimiento" faults was mapped by Eckels and others (1941); a somewhat larger area in the Atascadero quadrangle was mapped by Page (1972). Other investigations pertinent to this study are cited elsewhere in this report.

The cooperation of the many property owners of the Santa Margarita and Atascadero areas is gratefully acknowledged. Discussions with and encouragement by my colleagues are very much appreciated; they are too numerous to identify individually. Those who provided data, however, are cited elsewhere in the report. The manuscript was critically reviewed by Richard M. Stewart, Thomas E. Gay, Jr., and George B. Cleveland, all with the California Division of Mines and Geology.

ROCK UNITS

A variety of rock types is exposed in the map area, and 17 principal units (excluding subunits) are delineated on the geologic map (plate 1). Most of the units are sedimentary, although volcanic, igneous intrusive, metamorphic, and tectonic units are present. The ages, stratigraphic relations, and summary descriptions of the units are given on the composite stratigraphic columns (figures 3 and 4). Pre-Tertiary rocks on opposite sides of the Rinconada fault show great contrasts as to age and type, which are reflected both in the nomenclature and the organization of this report. The Cenozoic rocks show more subtle contrasts across the Rinconada fault, although the same formational names (Vaqueros, Monterey, Santa Margarita, Paso Robles) have been applied on both sides of the fault. Nonetheless, it should be kept in mind that the pre-Pliocene formations probably formed in widely separated areas and may since have been dislocated by as much as 30 or 40 miles.

PRE-TERTIARY ROCKS WEST OF RINCONADA FAULT
Franciscan Melange

The Franciscan melange of this report was previously mapped and designated as the San Luis Formation of the Franciscan Group by Fairbanks (1904). The San Luis term was later abandoned by the U.S. Geological Survey and replaced by the Franciscan Formation (Wilmarth, 1938, p. 1915). Recognizing the tectonic and chaotic nature of this unit, Hsü (1969) applied the term "melange". Although used in a less inclusive way by this writer, the term "Franciscan melange" adequately designates the chaotic rocks of the San Luis Obispo quadrangle. This unit is similar to the chaotic-type Franciscan found elsewhere in California (see Blake, 1970).

In the mapped area, the Franciscan melange is a chaotic mixture of sandstone and associated shale, altered volcanic rocks (greenstone), and lesser amounts of thin-bedded chert, serpentinite, diabase-gabbro, conglomerate, and blueschist facies metamorphic rocks. These rocks exist as small to large fault blocks and slices a few inches to half a mile long set in a highly sheared matrix of the same rock types and an undetermined amount of sheared shale. The relatively coherent blocks are also sheared and otherwise deformed internally but to a much lesser degree. Some of these blocks resemble rocks of the stratified upper Mesozoic unit mapped; others are dissimilar. The fault blocks and slices are typically exposed as "knockers"* surrounded by soil or landslide debris in a hummocky, grassy terrain. The unit is exposed only west of the Rinconada fault—mainly along the "Nacimiento" fault zone.

The Franciscan sandstone is a typically massive lithic graywacke with little or no bedding apparent. It is bluish or greenish gray where unweathered, medium to coarse grained, hard, and commonly cut by irregular veinlets of calcite and quartz. It weathers to tan or grayish brown. The sandstone

TIME—STRATIGRAPHIC (time) UNITS			APPROXIMATE AGE IN m y *
SYSTEM (period)	SERIES (epoch)	STAGE (age)	
Quaternary	Holocene		0.01
	Pleistocene		2-3
Tertiary	Pliocene		7-10
	Miocene	Delmontian	
		Mohnian	14
		Luisian	17
		Relizian	20.5
		Saucesian	24.5
	Oligocene	Zemorrian	
	Eocene		37-38
			53-54
	Paleocene		65
Cretaceous	Upper	Maestrichtian	
		Campanian	
		Santonian	82
		Coniacian	
		Turonian	
		Cenomanian	100
	Lower	Albian	
		Aptian	
		Barremian	118
		Hauterivian	
		Valanginian	
		Berriasian	136
Jurassic	Upper	Tithonian	146

Figure 2. Time-stratigraphic (time) units and approximate ages in millions of years as used in this report.

*A "knocker" is defined by Berkland and others (1973, p. 2296) as "a resistant rounded monolith, ranging from a few feet to several hundred feet in diameter, that generally stands out prominently above the level of the surrounding terrane".

Age		Symbol and Estimated Thickness (Feet)	Map Unit and Description
Quaternary	Holocene	Qyg 0-50	Younger alluvium - gravel, sand, mud; unconsolidated.
	Pleistocene	Qoo 0-200?	Older alluvium - conglomerate, sandstone, mudstone; weakly consolidated; includes relatively older (Qoa₁) and younger (Qoa₂) deposits locally.
	Pliocene E/or Pleistocene	TQp 0-400 TQpb 0-125	
Tertiary	Miocene	Tsm 300?- 2000	Paso Robles Formation - conglomerate, sandstone, mudstone; weakly to moderately consolidated stream and lake deposits. TQpb - basal gravel and coarse sand with large bored clasts of Tml and Tv.
			Santa Margarita Formation - fine to coarse, light gray, arkosic sandstone; includes conglomerate, siliceous mudstone and porcelanite, shell lenses, some diatomite; marine.
		Tmu 0?- 700 Tmls 0-300 Tmlv 0-1000 Tml 200-1400	Monterey Formation - well-bedded sequences marine upper siliceous (Tmu) and lower carbonate (Tml) beds; locally undifferentiated (Tml). Tmu - siliceous shale, mudstone, siltstone, porcelanite, chert, with diatomite, sandstone, tuff and minor bentonite and phosphatic beds. Tml - calcareous, foraminiferal, bituminous mudstone, siltstone with dolomite, shale, arkosic sandstone, siliceous beds, impure limestone; Tmls - friable, arkosic sandstone, Tmlv - olivene basalt.
	Miocene - Oligocene (?)	Tv 0-125 (350)	Vaqueros Formation - massive sandstone, conglomerate, impure bioclastic limestone; shallow marine.
		Tu 0-1000	Unnamed conglomerate - massive, poorly-bedded conglomerate, sedimentary breccia, sandstone; includes large boulders Ka sandstone; nonmarine.
Cretaceous	Late Cretaceous	— fault — Ka₄ 1000+ Ka₃ 1200 Ka 5200+ Ka₂ 1000 Kal 1000? Ka₁ 2000+ — Fault —	Atascadero Formation - undifferentiated sequences (Ka) massive biotitic sandstone with interbedded sandstone-mudstone turbidites; sparse marine fossils. locally subdivided near Paradise Valley: Ka₁ - dark mudstone with graywacke; Ka₂ - graywacke with dark mudstone, impure limestone; Ka₃ - poorly-bedded, massive sandstone with conglomerate, siltstone, mudstone; Ka₄ - massive, thick beds biotitic sandstone with thin-bedded sandstone-mudstone turbidites. Kal - undifferentiated, lower units of dark mudstone-shale with sandstone.
	Late Jurassic - Early Cretaceous	JKt 1500?+	Toro Formation - dark, brittle shale-mudstone with lithic graywacke and pebble conglomerate; locally siliceous with chert beds at base; marine fossils.
Jurassic	Late Jurassic?	v 1500?	Mafic volcanic rocks - greenstone and related altered lava (fine-grained to diabasic) and tuff breccia; crudely layered (stratified?); some pillows and vesicles; radiolarian chert locally; marine.
	Late Jurassic or older	sp-db 1300?	Ultramafic-mafic complex - massive and sheared serpentinite (sp) locally altered to silica-carbonate; cut by dikes and other intrusive masses of diabase, gabbro, pyroxenite, quartz diorite (db).
	Late Jurassic-Cretaceous	— Fault — f	Franciscan melange - chaotic mixture of fault blocks and slices of graywacke, shale and greenstone with subordinate chert, conglomerate, serpentinite, diabase-gabbro, and blueschist metamorphic rocks set in a sheared and shaly matrix; a tectonic unit.

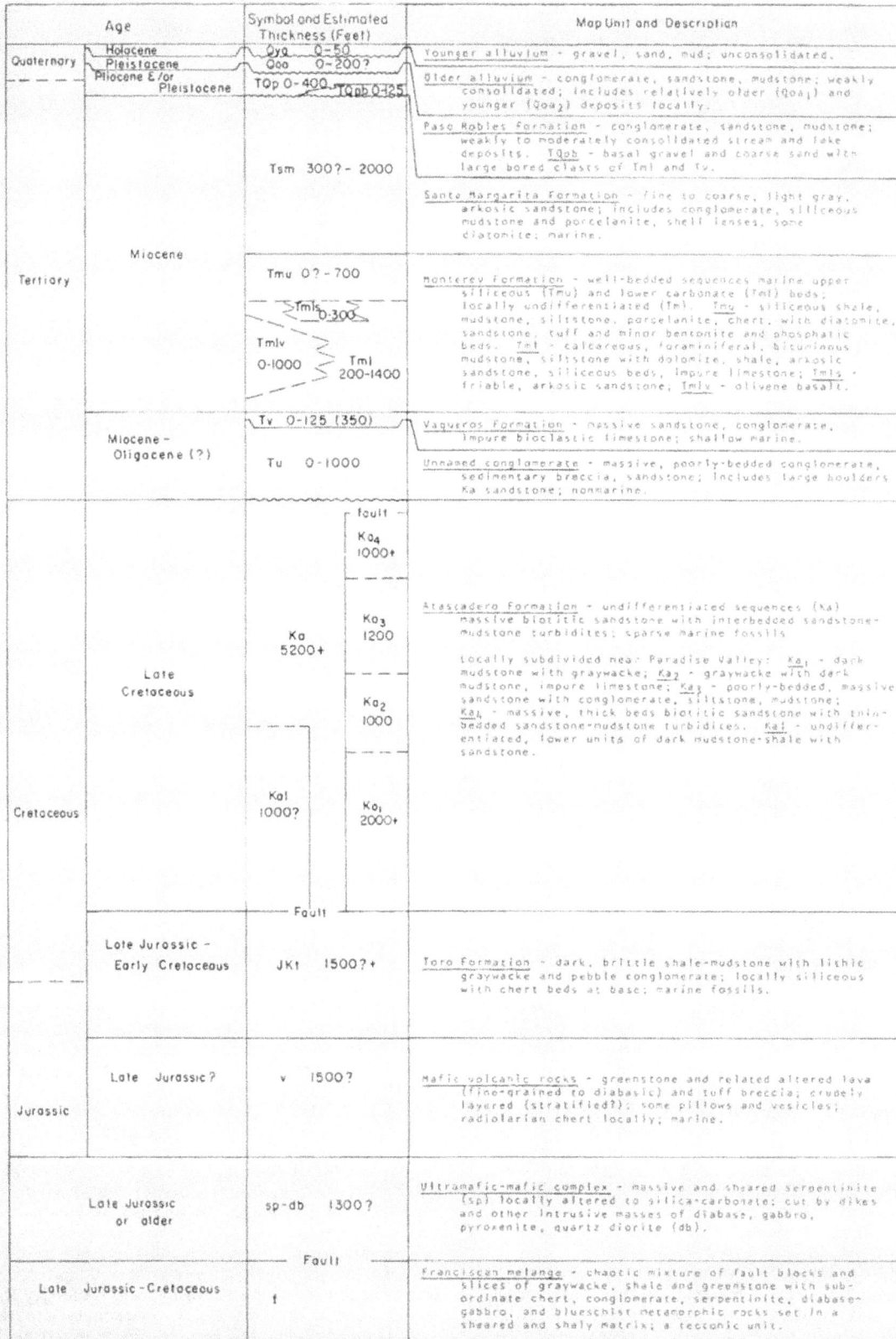

Figure 3. Composite stratigraphic column west of Rinconada fault.

Age		Symbol and Estimated Thickness(feet)		Map Unit and Description
Quaternary	Holocene	Qya	0-30	Younger alluvium - gravel, sand, mud; unconsolidated.
	Pleistocene	Qoa	0-50	Older alluvium - conglomerate, sandstone, mudstone, weakly consolidated; includes relatively older (Qoa₁) and younger (Qoa₂) deposits locally.
	Pliocene E/or Pleistocene	TQp	0-500	
Tertiary	Miocene	Tsm	100-200	Paso Robles Formation - conglomerate, sandstone, mudstone; stream and lake beds.
		Tm 900?	Tmu / --?--?--?-- / Tml	Santa Margarita Formation - arkosic sandstone, conglomerate, siltstone, coquina; nearly white to yellowish and reddish brown; shallow water marine. Monterey Formation - deformed sequence of undifferentiated marine beds (Tm) with upper siliceous (Tmu) and lower carbonate (Tml) members recognized but not mapped. Tmu - siliceous shale and siltstone, porcelanite, chert, diatomite, white sandstone, tuff. Tml - calcareous foraminiferal mudstone, siltstone interbedded with hard dolomite, porcelanite, arkosic sandstone and minor impure limestone and phosphatic beds.
	Miocene-Oligocene (?)	Tv	20-600? Tvc 0-500?	Vaqueros Formation - arkosic sandstone and conglomerate, nearly white to gray-brown; locally interbedded with siltstone and mudstone; marine. Nonmarine (?) conglomerate and sandstone (Tvc) at base near SE part of map.
	Oligocene (?)	Ts	0-800	Simmler Formation (?) - conglomerate with sandstone, mudstone; massive; poorly-bedded; weakly consolidated; unsorted granitic debris with boulders to 15' long.
Cretaceous	Late Cretaceous	Ku	4000+	Unnamed sandstone and conglomerate - massive to cross-bedded, medium to coarse, biotitic, arkosic sandstone and conglomerate sequences interbedded with turbidite sequences of well-bedded sandstone, siltstone, mudstone; minor limestone beds and concretions; marine and nonmarine (?).
	"Mid"-Cretaceous	Kgr		Granitic rocks - biotite granodiorite and quartz monzonite; uneven-grained; porphyritic; xenoliths present locally; cut by dikes of aplite, pegmatite, leucogranite.
	Pre-Late Cretaceous	m		Metamorphic rocks - dark gray schistose and gneissic rocks in small tabular pendant in Kgr.

Figure 4. Composite stratigraphic column east of Rinconada fault.

contains abundant quartz, altered volcanic debris, chert and plagioclase with lesser amounts of clastic sedimentary and metamorphic grains, degraded biotite, and various other (usually minor) constituents. K-feldspar is usually absent, although scattered grains (possibly altered volcanic glass in some cases), composing up to 0.5 percent of some samples, are present locally. The matrix consists of murky phyllosilicates, largely altered to chlorite and squashed unstable grains of mica and soft lithic fragments, which make up 10 to 20 percent of the rock. Calcite cement is sometimes present interstitially, in which case the framework grains are partly replaced by it. Dark gray, hard shale and siltstone are locally associated with the sandstone as thin interbeds. Shale and mudstone are rarely exposed and their abundance is uncertain but may constitute much of the sheared matrix of this unit.

Some sandstone within the Franciscan melange is atypical, being somewhat softer, brownish (weathered?), and locally constituting internally deformed stratified sequences with mudstone. This sandstone contains more matrix material (with less chlorite) and probably a higher percentage of altered and unstable grains. It superficially resembles Upper Cretaceous sandstone but contains no K-feldspar.

Rhythmically bedded sequences of graywacke, somewhat similar in lithology to the typical massive graywacke, were noted in several places. The largest of these is the small thrust plate designated as questionable Franciscan in SW 1/4 NE 1/4 sec. 10, T. 30 S., R. 13 E. An orderly, somewhat deformed sequence of 1/2-inch to 2-foot thick beds of lithic graywacke, with scattered pebbles and a few thin mudstone interbeds, is exposed here. Graded bedding and other turbidity current features are common. The sequence is in sharp contrast with the chaotic Franciscan melange to the northwest and northeast. To the southeast, the sequence appears to become progressively more deformed and crushed, and it grades into the more chaotic melange. Rocks of this and other Franciscan sequences are rather similar to well-bedded sandstones mapped as Toro

Formation (partly queried on map). The distinctions between Franciscan and Toro are based largely on structural relationships.

Greenstone is a general field term applied to the abundant mafic igneous rocks of presumed extrusive origin. Although alteration (mostly to green chlorite) and weathering (indicated by reddish-brown coloration) is extensive, diabasic textures and pillow structures can be seen locally. For the most part, the greenstones are sheared and show little in the way of primary textures, structures and stratification. Larger and more intact sequences of mafic volcanic rock were mapped separately from Franciscan (see under Mafic Volcanic Rocks).

Chert is common as fault blocks of rhythmically bedded, contorted bands 1 to 2 inches thick. It is usually greenish gray or reddish brown but may be other colors. Thin partings of shale commonly separate the chert layers, especially in the reddish-brown varieties. It consists predominately of finely crystalline quartz and chalcedony, including sparse to abundant radiolarians. Small but variable amounts of clay and other phyllosilicates are present as impurities. Quartz veinlets filling closely spaced fractures are characteristic. The chert is commonly brecciated and recrystallized. In one place, it is manganiferous (deposit 2, table 8).

Well-indurated pebble to cobble conglomerate is present locally. These contain clasts of chert, volcanic rocks (greenstone and other metavolcanics), some clastic sedimentary debris (argillite, siltstone, sandstone) and quartz. Other types of clasts may be present, but no silicic granitic debris was seen. The conglomerate is generally similar to that of the Toro Formation.

The metamorphic rocks are not abundant and consist of hard blueschist facies "knockers" of varying composition. These are usually schists with contorted layers composed of soda amphibole, actinolite, quartz, calcite, lawsonite, chlorite, albite, and other minerals in various combinations. Yellow-brown and reddish jasper locally is associated with soda amphibole schist.

Minor amounts of serpentinite, diabase, gabbro, and related rocks identical to rocks of the ultramafic-mafic complex (described below) are present locally. These rock types are included with the Franciscan only where inclusions are too small or too chaotically mixed with other rock types to portray on the map.

The Franciscan melange appears to be in fault contact with all other Mesozoic rock units. The oldest rock unit to rest depositionally on the Franciscan is the Monterey Formation, basal beds of which are of early Miocene age. The unnamed conglomerate (Oligocene and/or early Miocene) in the northwest part of the area also may be depositional on the Franciscan, but the contact is not clearly exposed.

The tectonic origin of the Franciscan is indicated by its chaotic content and pervasively sheared matrix. Relatively coherent masses of Toro Formation, mafic volcanic rocks, and serpentinite (only partly delineated on map) in places appear to grade tectonically into the chaotic Franciscan along the structural strike. Such tectonic mixing is believed to have taken place largely prior to latest Cretaceous time as the K-feldspar-rich Late Cretaceous strata are less intimately incorporated into the Franciscan than the older Mesozoic units. In fact, typical massive graywacke of the Franciscan contains little or no K-feldspar, which suggests that it is probably no younger than mid-Cretaceous (Hart, 1971). Intense Cenozoic deformation has overprinted and partly obscured the earlier deformation and melange relations. Monterey Formation rocks, for example, are found locally in the melange but occur mainly along linear fault zones.

Based on current concepts of plate tectonics, it seems likely that the Franciscan melange was formed within a late Mesozoic subduction zone between converging oceanic and continental plates, incorporating elements of both plates. Subduction and attendant melange-forming processes presumably abated by late Oligocene time as the Franciscan is overlain by and has contributed debris to the Vaqueros and basal Monterey Formations within and near the mapped area (see Page, 1970b).

Ultramafic-Mafic Complex

This unit consists of serpentinite and intrusive bodies of altered diabase and related rocks. These rocks are exposed discontinuously along the southwest margin of the Franciscan melange and as small fault slices (mostly serpentinite) within the melange. Part of this complex was included with the Cuesta Diabase by Fairbanks (1904).

The massive to sheared greenish serpentinite is derived from various olivine- and pyroxene-bearing ultramafic rocks. In the Franciscan melange, it occurs mainly as thin fault slices that tend to align along rather distinct faults. Somewhat larger masses of serpentinite and associated mafic and ultramafic intrusives at Eagle Peak, Frog Pond Mountain, and the area northwest of Devils Gap tend to lie at the base of thrust plates, particularly where overlain by mafic volcanic rocks. The margins of the serpentinite masses are nearly everywhere sheared, often with the development of talc and actinolite rocks. Alteration to silica-carbonate rock also is common locally, especially southeast of Santa Margarita. Veinlets of chrysotile asbestos are most common in the large thrust plate of serpentinite northwest of Burrito Creek. Here, the veinlets are as much as half an inch wide and in places may constitute up to 10 percent of the rock.

Hard, dense, green-gray to dark-gray diabase is complexly intermingled with the serpentinite. The diabase consists of fine to very coarse plagioclase and pyroxene (augite?) that is largely relict and replaced by saussurite, actinolite, chlorite, and other alteration products. Exposures are generally poor due to the development of deep reddish soil and the dense brush that grows on it. Where exposed in a road cut east of Frog Pond Mountain (SW 1/4 sec. 3, T. 28 S., R. 12 E.), the serpentinite is intruded by west-trending, nearly vertical diabase dikes to several feet thick. Similar mafic dikes, partly altered to rodingite (Hsu, 1969, p. 34), are better exposed along State Route 41 just west of the mapped area.

Various other rocks of mafic to ultramafic composition, observed in association with diabase, include hornblende gabbro, pyroxenite, pyroxene-bearing peridotite (harzburgite?), and leucocratic quartz diorite. Primary minerals and their alteration products include one or more pyroxenes, plagioclase, chlorite, saussurite, actinolite, olivine, serpentine minerals, and magnetite. Most or all of these rocks presumably intrude the serpentinite as dikes and sills.

The contact of the ultramafic-mafic complex with the overlying mafic volcanic rocks on Frog Pond Mountain is believed to be essentially normal (i.e., depositional and/or intrusive), although in detail it is locally somewhat sheared. Contacts with the other Mesozoic rocks appear to be faulted everywhere.

The age of the serpentinite is uncertain, but it may be the oldest rock type west of Rinconada fault. The intrusive diabase and associated gabbro and peridotite presumably are genetically related to the overlying mafic volcanic rocks of probable Late Jurassic age. A Late Jurassic age was determined radiometrically for gabbro from mafic-ultramafic sequences in two places below the Great Valley sequence east of the San Andreas fault (Lanphere, 1971).

This unit, together with the overlying mafic volcanic rocks and bedded chert, probably represents remnants of former upper mantle and lower oceanic crust (ultramafic-mafic complex), upper oceanic crust (mafic volcanic rocks) and overlying abyssal ooze (radiolarian chert). Such sequences, commonly termed ophiolites, have been identified near the map area and elsewhere in the California Coast Ranges (Page, 1972; Bailey and others, 1970).

Mafic Volcanic Rocks

Relatively large masses of altered mafic volcanic rocks—commonly known as green-stone—are exposed at Eagle Peak, Frog Pond Mountain, and Devils Gap. Although faulted and discontinuous, these rocks may represent an intermediate stratigraphic unit about 1500 feet thick between the ultramafic-mafic complex (below) and the Toro Formation (above) (see Page, 1972). These volcanic rocks were designated (in part) as the Cuesta Diabase by Fairbanks (1904), who considered that unit to be intrusive. Although the greenstone presumably includes some dikes or sills, the unit displays common volcanic features. Similar mafic volcanic rocks occur as tectonic blocks within the Franciscan melange. Four of these blocks—near Paradise Valley and east of Eagle Peak—were sufficiently large and coherent to map separately from the Franciscan. Small exposures of thin-bedded chert are included locally, both within and at the top

of the mafic volcanic unit. Thin interleaves of Toro-like shale within the volcanic units are believed to be fault slices. Steep topography and dense vegetation hamper observation of the mafic volcanic unit.

The mafic volcanic rocks erode differentially, forming resistant outcrops that suggest crude stratification or layering. These rocks are typical greenstones, being greenish gray when fresh but weathering readily to reddish-brown outcrops and soil. Most of the rocks are basalt and diabase, and some have vesicular textures. Pillow structures can be seen rarely but are difficult to identify in these altered and partly deformed rocks. Associated and presumably interbedded with these lavas are basalt porphyry, mafic tuff, and crystal tuff breccia. Excellent exposures of these mafic rocks can be seen in Graves Creek just east of Paradise Valley where tuff fragments reach at least 4 inches across and altered plagioclase phenocrysts and xenocrysts an inch in diameter. Altered diabase and tuff also are well exposed in road cuts at Devils Gap.

Petrologic characteristics of these rocks are variable, partly because of greenschist facies metamorphism. Recognizable primary minerals of the basalt and diabase are plagioclase (laths, phenocrysts), relict augite and related pyroxenes, and accessory black opaques. These are partly to completely altered to varied mixtures of chlorite, albite, calcite, epidote-clinozoisite (including saussurite), sericite, quartz, and nontronite. Vesicles are commonly filled with calcite and, less often, nontronite, chalcedony, chlorite, and actinolite(?). The crystal tuff breccia contains altered plagioclase xenocrysts and murky fragments of altered glass(?) set in vesicular nontronite (altered glass) with scattered small crystals of quartz and feldspar(?).

Small outcrops of thin-bedded chert and locally associated siliceous tuff(?) are included with the mafic volcanic unit. Most outcrops occur along the southwest margins of the volcanic units and probably overlie these units. The few occurrences of chert within the volcanic units may be tectonic inclusions. The chert is relatively pure, brittle, hard, fractured and veined, and locally brecciated. It is colored various shades of cream, blue, green, gray, and reddish brown; contains abundant radiolarians; and is similar to chert of the Franciscan melange. Pale greenish-gray siliceous tuff(?) is locally associated with the chert half a mile northwest of Devils Gap. The chert exposures are partly obscured by talus debris. Small irregular masses of reddish jasper formed as secondary deposits within the mafic volcanic sequence.

The stratigraphic relations and age of the mafic volcanic rocks are difficult to establish within the map area, because many of its contacts with other Mesozoic units are faults and well-exposed contacts are rare. The depositional relationship of this unit to the underlying ultramafic-mafic complex in the Frog Pond Mountain-Devils Gap area is indicated by: 1) the approximate conformability between the crudely layered volcanic rocks and the southwest-dipping basal contact of the unit; 2) the position of the ultramafic-mafic complex below the mafic volcanic units; 3) the diabase dikes which cut the serpentinite and which may have acted as feeders to the mafic volcanic rocks. The mafic volcanic rocks appear to be overlain conformably by thin-bedded chert, which may be gradational with the overlying siliceous beds of the lower Toro Formation(?) along Atascadero Creek. This relationship suggests that the mafic volcanic rocks of the Frog Pond Mountain-Devils Gap area are no older than the latest Jurassic. Based on stratigraphic, structural, and fossil evidence, Page (1972) interprets these volcanic units and similar units to the south and west to be late Jurassic in age.

If the supposed relations are correct, the mafic volcanic rocks of the Eagle Peak-Devils Gap exposures probably formed as a result of submarine eruptions along a spreading oceanic rise. The radiolarian chert presumably represents former pelagic ooze deposited on the volcanic rock as it migrated away from the rise. The lack of limestone suggests deposition below the calcium carbonate compensation depth (i.e., below 12,000-15,000 feet). Whether the volcanic units just west of Paradise Valley and east of Eagle Peak formed in the same way, or even at the same time, is uncertain, although they are at least partly of submarine origin and are lithologically similar to the volcanic rocks to the southwest.

The mafic volcanic unit of the Eagle Peak-Devils Gap area is correlative in age, general lithology, and stratigraphic position to other mafic volcanic sequences that lie beneath the Great Valley sequence elsewhere in the California Coast Ranges (Bailey and others, 1970).

Toro Formation

The Toro Formation consists of dark shale and thinly interbedded lithic sandstone and pebble conglomerate of Late Jurassic and Early Cretaceous age. It is exposed as deformed and disrupted sequences along the southwest margin of the map area where the unit is 1000 to 1500 feet thick or more. These outcrops comprise part of the northeast margin of a 20-mile long northwest-trending synform of contiguous Toro Formation outcrops (Fairbanks, 1904; Jennings, 1958). Topographically, the Toro is characterized by a highly dissected terrain covered with brush or grass. Landslides are common on steep slopes.

H. W. Fairbanks (1904) named the formation for exposures in Toro Creek, 1 or 2 miles west of the mapped area. He considered the Toro to be the "local representative of the Knoxville group". The unit name was abandoned as unnecessary by the U.S. Geological Survey (Wilmarth, 1938, p. 2169) and replaced by the Knoxville Formation. Later, rocks of the unit were identified, in whole or part, as the Marmolejo Formation (Taliaferro, 1944) and as the Great Valley sequence (Bailey and others, 1964; Gilbert and Dickinson, 1970). Of the several designations, the name Toro seems most ap-

propriate to the purposes of this report. However, since the Toro has no clearly definable stratigraphic top, it is convenient to view the unit as the lowest part of an undefined and incomplete "Great Valley-type sequence".

Typically, the Toro Formation consists of dark-gray to olive-gray, brittle shale and mudstone with thin interbeds of fine graywacke and siltstone and scattered limy concretions. Massive thick beds or lenses of medium- to coarse-grained, lithic graywacke and pebble conglomerate are sporadically distributed through the shales; and these also characterize the unit. The sandstone is a lithic wacke composed of abundant volcanic debris (partly mafic), sedimentary rocks (chert plus shale, siltstone, sandstone) and subangular quartz, with subordinate amounts of plagioclase, metamorphic rock grains and micas. K-feldspar is virtually absent. The sand grains are densely packed. Murky phyllosilicates, including secondary chlorite, squashed unstable grains and some calcite constitute most of the intergranular material. Turbidity current features (graded beds, cross-lamination, rhythmic interbedding with shale) are abundant. The conglomerate contains abundant rounded, dark-gray pebbles of chert, aphanitic volcanic rocks and other rock types; granitic clasts are absent. Southeast of Eagle Peak and 2000 feet northwest of Devils Gap, near the ultramafic-mafic contacts, the shale is hard, siliceous, and partly tuffaceous(?). Silicified shale with black chert also is exposed near the base of the Toro Formation(?) in Atascadero Creek west of Eagle Peak. These impure cherts and siliceous shales were mapped as an unnamed bedded chert and partly as Toro Formation by Page (1972).

Atypical gray to greenish-gray, fine- to medium-grained, laminated to well-bedded graywacke (turbidite) with mudstone interbeds is exposed southeast of Eagle Peak in a deformed sequence. These beds contain 1-5% K-feldspar and are somewhat similar to the lower beds (Kal and Ka₁) of the Atascadero Formation. Other predominantly sandy somewhat atypical sequences, tentatively identified as Toro, are exposed west of Eagle Peak in Atascadero Creek and west of the large serpentinite mass in sec. 11, T. 30 S., R. 13 E. These sequences contain sedimentary breccias of light greenish-gray chert and some serpentinite locally and may represent basal Toro beds.

Internal deformation apparently is widespread within the Toro Formation, as indicated by the abundant and sudden variations in bedding attitude. The beds commonly are extensionally sheared and fractured, ductilely deformed, and contorted. Shear-type deformation also can be seen in some sandstones, both in thin-section and at the outcrop. The fact that shearing tends to be pervasive suggests gravitational sliding or slumping as a major cause of deformation. This may have been either soft- or hard-rock deformation, depending on depth of burial (i.e., load; see Kehle, 1970). Strained and broken sand grains indicate some of the deformation to be of the hard-rock type.

The top and much of the base of the Toro Formation appear to be missing, the formation generally being in fault contact with the other upper Mesozoic units. Locally, however, the base appears to be depositional on the mafic volcanic unit. For example, in Atascadero Creek west of Eagle Peak, a deformed sequence of Toro Formation(?) sandstone and shale concordantly overlies a mafic volcanic sequence. Here, the base of the sedimentary sequence contains siliceous shale and thin interbeds of black chert. This overlies a thin (10-20 feet thick?) light-colored chert or siliceous tuff(?) unit that rests on mafic volcanic rocks. Although a waterfall in the creek limits access and direct observations, apparent transitional relations between terrigenous sediment and chert suggest that the Toro(?) sequence may rest depositionally on the chert and mafic volcanic rocks. Cherty radiolarian shale resting conformably on mafic volcanics has been identified more extensively within and adjacent to the map area by Page (1972).

Megafossils, mainly species of *Buchia*, indicate the age of the Toro Formation to be Tithonian (latest Jurassic) to Valanginian (Early Cretaceous) (table 1). However, the presence of 1-5% K-feldspar in deformed beds, 1 mile southeast of Eagle Peak, and general petrologic similarity of these beds to unit 1 of the Paradise Valley sequence and other lower beds of the Atascadero Formation suggest that part of the Toro may be mid-Cretaceous in age (Hart, 1971). Upper Cretaceous palynomorphs at two localities west and northwest of the mapped area (Gilbert and Dickinson, 1970) also suggest that the Toro includes some beds of post-Valanginian age. The lack of upper Lower Cretaceous fossils here and elsewhere in the Santa Lucia Range may indicate a major stratigraphic gap within the Toro and equivalent strata.

Table 1. Megafossils of the Toro Formation.

Map locality number	Location (Sec-T-R)	Key fossils	Age
SLO 145[1]	11-30S-13E	Belemnite (? *Acroteuthis* sp.)	Early Cretaceous (?)
SLO 223[2]	33-29S-13E	*Buchia piochii* (Gabb)	Late Jurassic (Tithonian)
SLO 293[2]	10-29S-12E	*Buchia pacifica* Jeletzky	Early Cretaceous (Valanginian)
SLO 301[2]	25-28S-11E	*Buchia piochii* (Gabb) or *B. uncitoides* (Pavlow)—specimens crushed	Late Jurassic (Tithonian) or Early Cretaceous (Berriasian)
SLO 307[2]	32-28S-12E	*Buchia pacifica* Jeletzky and *B. keyserlingi* (Lahusen)	Early Cretaceous (Valanginian)

[1] Identified by J. Wyatt Durham, University of California at Berkeley, March 1968.
[2] Identified by David L. Jones, U.S. Geological Survey, Menlo Park, California, January 1971.
   Same as USGS Mesozoic (M) locality numbers as follows:
      SLO 223=M5521   SLO 301=M5536
      SLO 293=M5535   SLO 307=M5537

The presence of marine fossils, dark shale, and turbidites indicate that the Toro sediments were deposited in a marine basin adjacent to a landmass. Terrigenous constituents in typical Toro sandstone and conglomerate reflect a mixed volcanic, sedimentary and low-grade metamorphic provenance. The atypical rocks, which may be somewhat younger, indicate additional acid plutonic detritus (K-feldspar, biotite). Varying sediment compositions, internal deformation, apparently missing strata within and at the top of the Toro, and relationships to other Upper Mesozoic units suggest that the formation was deposited in a tectonically active environment that affected both the basin of deposition and the provenance area. The Toro may represent a transition from abyssal pelagic deposition of radiolarian ooze on oceanic volcanic rocks to terrigenous sedimentation within a developing continental margin—presumably between a consuming trench and a related magmatic arc.

The Toro Formation is similar in age and lithology to the lower part of the Great Valley sequence east of the San Andreas fault. In the Santa Lucia Range, it is partly or wholly equivalent in age to the Marmolejo and Knoxville Formations of Taliaferro (1943; 1944), the Jollo Formation and part of the "Franciscan group" of Hall and Corbato' (1967), and the Jollo and Knoxville Formations of Brown (1968).

Atascadero Formation

The Atascadero Formation of this report is used to designate the loosely correlated, sparsely fossiliferous, disrupted Upper Cretaceous sequences of four separate fault blocks. The unit is virtually the same as the Atascadero Formation of Fairbanks (1904) with the principal exception of beds in the Paradise Valley area that were designated as Toro Formation by Fairbanks. The Atascadero, named for outcrops in Atascadero Creek, was believed by Fairbanks to be unconformably depositional on the Toro Formation. Although the Atascadero was abandoned by the U.S. Geological Survey in favor of the Chico Formation (Wilmarth, 1938, p. 159) and later included with the Asuncion Formation of Taliaferro (1944), the original name probably is the most suitable for the Upper Cretaceous rocks of the map area. Informally, the Atascadero has been included with the upper Mesozoic Great Valley sequence (Bailey and others, 1964; Bailey and others, 1970; Gilbert and Dickinson, 1970).

In the mapped area the Atascadero Formation is widely exposed in the Santa Lucia Range and also is present in the subsurface northeast of there. It lies entirely west of the Rinconada fault. It commonly is identified by bold outcrops of massive thick-bedded coarse sandstone and some conglomerate that weather to yellowish or reddish brown. Creek and road cut exposures, however, reveal abundant well-bedded sandstone, siltstone, and mudstone. Internal disruption and deformation of the unit is common, and stratigraphic markers can be traced only for short distances. Four subdivisions (units 1 to 4) of the Atascadero were recognized and mapped locally in the Paradise Valley area. Dark mudstone-sandstone units in the lower part of other Atascadero fault block sequences also were identified and partly mapped. For the most part, the Atascadero Formation consists of undifferentiated sedimentary sequences.

The Atascadero Formation is characterized by sequences of light- to medium-gray or olive-gray, thin- to thick-bedded sandstone with interbedded siltstone, mudstone, and subordinate amounts of conglomerate and impure limestone. Bedding varies from massive to laminated, occasionally revealing large-scale cross-bedding. Turbidity current and other strong current features are common and include rhythmic bedding, graded beds, cross lamination, scour and fill, convolute bedding, intraformational clasts, and various sole markings. Paleocurrent directions near Atascadero are reported to be toward the west and southwest (McClure, 1969). Internal deformation is widespread and locally intense. This includes: 1) pervasive shears in sandstone subparallel to the bedding; 2) pinched-off sandstone beds, including occasional boudins; and 3) pinching and swelling of mudstone beds. Some of the deformation may be of "soft rock" type (e.g., slumping and sliding contemporaneous with deposition), but much of it is "hard rock" (as indicated by common microscopic shears and deformed grains in sandstone) and probably is the result of large-scale overthrusting or gravity sliding.

Typical sandstones of the Atascadero consist of fine to very coarse and pebbly, well-packed grains of quartz (30-40%), feldspars (30-50%), volcanic and other lithic debris (10-30%), and biotite (2-10%). Abundant K-feldspar (10-30%) and crinkly biotite characterize the rock. Interstitial material includes phyllosilicates, squashed unstable grains of biotite and soft lithic fragments, and calcite. Alteration is common; secondary chlorite is weakly developed in most sandstone and laumontite was noted in several samples. In the field, the sandstones are classed both as arkose and graywacke, but most often resemble a "dirty" arkose.

There are lenses of conglomerate, perhaps as much as 40-50 feet thick. Some sandstone also is conglomeratic. Clasts generally are rounded cobbles of volcanic rocks with subordinate granite and quartzite. Clast counts of conglomerate at three localities along State Route 41 were reported by McClure (1969) to be silicic volcanics (35%), basic volcanics (28%), quartzite (14%), plutonic igneous (12%), vein quartz (4%), other (7%). Similar clasts in about the same proportions are noted elsewhere in the Atascadero conglomerates.

Mudstone associated with the typical feldspathic sandstone tends to be soft and generally nonfissile. It weathers conchoidally, eventually breaking into small chips.

In several places in three of the four fault blocks, lower beds of dark mudstone and sandstone

can be mapped with varying degrees of confidence and completeness in the Atascadero Formation. These are identified as Kal on the map, except in the Paradise Valley area where a transitional sequence of lower dark mudstone (Ka_1) grades upward into typical massive feldspathic sandstone (Ka_4) (see below). The lower beds are characterized by hard to firm, dark-gray to olive-gray mudstone with thin interbeds and concretions of impure limestone. The sandstone contains abundant lithic debris and generally 0.5-10% K-feldspar. As such, the lower units are intermediate in character between typical Toro and Atascadero rocks. The lower beds also show considerable internal deformation. Although the lower dark mudstone sequences are generally similar to each other throughout the map area, stratigraphic relations among the sequences are uncertain.

An apparently intact, transitional, 5000-foot sequence of lower dark mudstone and upper typical massive sandstone of the Atascadero Formation is exposed in and near Paradise Valley. The sequence is informally referred to herein as the Paradise Valley sequence. It is subdivided into units 1 to 4 from the base upwards (Ka_1 to Ka_4 on map). Although the sequence shows minor structural complications, the units are conformable and partly gradational with each other. However, the sequence becomes indistinct toward the southeast and appears to grade into deformed undifferentiated rocks largely typical of the Atascadero. Unit 1 of the sequence consists of dark mudstone with thin to thick interbeds of well-bedded graywacke, siltstone, and minor limestone lenses. Ten samples of graywacke contained 0.5 to 10 percent K-feldspar with a mean of 4.6 percent (Hart, 1971 and "K-feldspar" in press). Unit 1 grades upward into unit 2, which mainly consists of flaggy to massive, calcareous graywacke (6-15% K-feldspar) with interbedded dark mudstone and impure limestone. Unit 3 conformably overlies unit 2, but the contact is locally faulted and obscure. A basal cobble conglomerate is locally exposed. Unit 3 is relatively soft, poorly bedded medium-gray to tan sandstone with interbedded mudstone, siltstone, and local conglomerate lenses. The sandstone is intermediate in character (abundant volcanic debris and 8-15% K-feldspar) between that of units 2 and 4. Unit 4 conformably overlies unit 3 and contains abundant massive tan-weathering coarse feldspathic and biotitic sandstone (20-25% K-feldspar) with interbedded sequences of finer sandstone, siltstone, and mudstone.

Aside from the Paradise Valley sequence, other sequences in the Atascadero Formation are so disrupted and difficult to correlate that a composite thickness cannot be determined. However, the aggregate thickness probably is two or three times the exposed thickness of 5000+ feet for the Paradise Valley sequence. It is likely that the several allochthonous Atascadero sequences were parts of a common Upper Cretaceous continental margin sequence, but their initial depositional relations are largely undetermined. All of the contacts between the Atascadero and the other upper

Mesozoic rock units appear to be faults. Nowhere was the Atascadero Formation seen to be resting depositionally on older rocks.

The few fossils found in the Atascadero Formation are of Late Cretaceous age (table 2). Fossils from localities associated with typical massive sandstone of the fault block southwest of the "Nacimiento" fault zone indicate a latest Cretaceous (Campanian-Maestrichtian) age for much of the formation. Additional mollusk localities near SLO 207 and 291, 2 to 4 miles easterly of Eagle Peak, also indicate a Late Cretaceous age (Page, 1971, personal communications; Gilbert and Dickinson, 1970). The only other diagnostic fossils found in the Atascadero Formation are from a shelly mudstone bed near the base of the lowest unit of the Paradise Valley sequence (SLO 247 in table 2). These indicate an early Late Cretaceous age. On the basis of sparse fossil evidence and gross lithologic characteristic—including K-feldspar content (see Hart, 1971)—the Atascadero Formation appears to represent the disrupted remnants of a mid-Cretaceous to latest Cretaceous sedimentary sequence with an undetermined top and base. Whether such a sequence was continuous and complete is uncertain. The upper parts of the unit correspond in lithology and partly in age to the Asuncion Formation to the northwest (Taliaferro, 1944) and the Carrie Creek Formation to the southeast (Hall and Corbato', 1967). The lower part of the Atascadero Formation is partly correlative with the Jack Creek Formation to the northwest (Taliaferro, 1944) and to the Adobe Flat Member of the Panoche Formation east of the San Andreas fault (Maddock, 1964; Bishop, 1970).

Table 2. Fossils of the Atascadero Formation.

Map locality number	Location (Sec-T-R)	Key fossils	Age
MOLLUSKS SLO 147[1]	14-30S-13E	Meekia daileyi Saul and Popenoe	Late Campanian or early Maestrichtian
SLO 207[2]	11-29S-12E	Glycymeris veatchii (Gabb) var. anae Smith— giant specimens	Very late Cretaceous (Campanian or Maestrichtian)
SLO 247[2]	19-28S-12E	Glycymeris sp.; Linearia multicostata (Gabb)	Late Cretaceous— probably Cenomanian or Turonian
SLO 291[2]	18-29S-13E	Baculites sp.	Late Cretaceous
PALYNO-MORPHS SLO 3[3]	20-28S-12E	Angiosperm pollen (2 types);Dinoflagellate (1 type)	Late Cretaceous(?)

[1] Identified by W. P. Popenoe and L. R. Saul, University of California at Los Angeles.
[2] Identified by D. L. Jones, U.S. Geological Survey, Menlo Park.
[3] Identified by W. R. Evitt, Stanford University.

Based on the widespread occurrence of the turbidites, presence of marine fossils, paleocurrent directions, and abundance of terrigenous clastic sediments, the Atascadero Formation must have been deposited in one or more subbasins along a continental slope west or southwest of a large landmass. A general upward increase in amount and coarseness of sand suggests progressive uplift of the bordering landmass and/or progradation westward of the Atascadero sedimentary wedge. Progressive upward changes in sediment character within the several Atascadero fault blocks indicate a provenance in which 1) the volcanic rocks changed from mafic to predominantly silicic and 2) the acid plutonic rocks were either progressively unroofed or became more acidic with successive intrusions (see Hart, 1971, and "K-feldspar" in press).

PRE-TERTIARY ROCKS EAST OF RINCONADA FAULT
Metamorphic Rocks

A small pendant of dark-gray foliated metamorphic rocks is exposed between the Middle and East Branches of Huerhuero Creek in the northeast part of the map area. The pendant, which is steeply dipping and less than 20 feet thick, is enclosed in granitic rocks and both are cut by irregular dikes of pegmatite and aplite. The metamorphic rocks are weakly schistose to gneissic and show the effects of contact metamorphism. They are composed of anhedral hornblende, two generations of plagioclase, and partly chloritized biotite with small amounts of K-feldspar, sphene, black opaques, and minor amounts of apatite and garnet. Weathered schistose xenoliths of similar composition are distributed unevenly throughout the granitic rocks. Because primary textures and minerals are not preserved in the metamorphic rocks, their nature prior to metamorphism is unknown. However, marble, schist, and quartzite reported elsewhere in the La Panza Range (Compton, 1966) show that some sedimentary rocks previously existed in the region. It is presumed that metamorphism occurred during the Cretaceous emplacement of acid plutonic rocks.

Granitic Rocks

Cretaceous granitic rocks are widely exposed in the map area as a western extension of the La Panza Range. The granitic rocks are confined, in surface outcrops and apparently in the subsurface, to the area east of the Rinconada fault. They are deeply weathered and commonly decomposed, giving rise to a highly dissected terrain that supports a dense growth of sagebrush. Unweathered granitic rock is exposed only at the Roselip (Kaiser) quarry. Slightly weathered rock is widely but intermittently exposed along the Salinas River and in stream bottoms. The great bulk of this rock unit consists of biotitic granodiorite and adamellite in almost equal amounts. These are cut by large dikes of leucogranite (biotitic aplite) and abundant small resistant dikes of aplite and pegmatite. Compositional gradations from aplite to leucogranite to adamellite and granodiorite suggest that all of these rocks formed during a single intrusive event or a closely related sequence of events.

The unweathered granodiorite and adamellite are light to medium gray with pink or green casts. They weather readily to yellow-brown colors. Most rocks are coarse grained, porphyritic, hypidiomorphic, and seriate, although leucocratic varieties tend to be finer and more even grained. Modal analyses of 19 feldspar-stained samples indicate 11 samples to be granodiorite and 8 to be adamellite (figure 5). Over-all compositional ranges (and means) of these slabs, in percent, are as follows: Quartz 23.8-39.1 (30.2); plagioclase 32.5-48.9 (41.0); K-feldspars 11.1-32.3 (20.4); biotite (and other micas) 2.6-12.3 (7.7); minor accessories 0-1.3 (0.6).

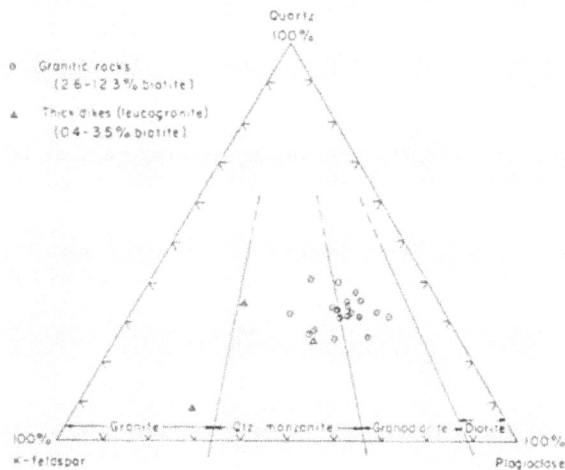

Figure 5. Ternary diagram showing volumetric compositions of 22 samples of granitic rock, western La Panza Range. Compositions are based on modal counts (300 minimum) of feldspar-stained slabs.

Additional detail was revealed in seven thin sections. Quartz often is present as a mosaic of sutured intergrowths, and myrmekite is present in all samples. The plagioclase is subhedral oligoclase and sodic andesine which is commonly zoned. The K-feldspar is largely untwinned orthoclase. Phenocrysts of K-feldspar, as much as an inch long, contain abundant partly aligned inclusions of plagioclase. Less often, the plagioclase contains small K-feldspar inclusions. Inclusions, such as apatite and biotite in quartz, and of quartz and biotite in K-feldspar, are common. Perthitic textures are present but not highly developed. Biotite is strongly pleochroic and partly chloritized. Minor accessories noted are sphene, apatite, ilmenite (partly altered to leucoxene), magnetite, muscovite, and hematite. Complex intergrowths and the abundance of inclusions suggest that the granitic rocks were ultimately emplaced as crystal mush after at least one earlier period of crystallization. These rocks are typical of the granitic rocks exposed throughout the La Panza Range (Compton, 1966; Ross, 1972).

Dike rocks are abundant and represent various late stages of emplacement. Leucocratic dikes of adamellite and aplite (0.4-3.5% biotite) reach thicknesses of as much as 30 feet or more. Thin dikes of aplite, pegmatite, and some graphic granite tend to occur in swarms and probably are late fracture fillings. Most of these dikes are 1 to 6 inches thick. They are composed of the same mineral species, in varying proportions, as the granitic rocks (figure 5).

The granitic mass is widely broken by faults, shears, and fractures that show a variety of orientations and types of movement (horizontal, oblique, and vertical). Most of these structures are minor and serve mainly to shatter the mass. The over-all patterns are not clear and probably represent variously oriented secondary stresses. Most of the fractures are presumed to be manifestations of a regional right-lateral shear imposed by the flanking Rinconada fault in the mapped area and the Huerhuero fault east of the area (see Jennings, 1958).

The age of the granitic rocks, based on the overlying unnamed Upper Cretaceous rocks to the south, is pre-latest Cretaceous (i.e., pre-late Campanian), or about 75 million years or older. A single K-Ar age determined for biotite indicates a minimum age of 80.2 m.y. (Curtis and others, 1958; recalculated by Compton, 1966). Additional K-Ar

Table 3.   Potassium-argon ages of biotite (partly chloritized) from granitic rocks of the western La Panza Range.

Sample number	Locality[1] (Sec-T-R)	Rock type	Percent K	Ar^{40} rad. ppm	$\dfrac{Ar^{40} \text{ rad.}}{Ar^{40} \text{ total}}$	Apparent age[2] (million years)
SLO 48*	6-30S-14E	Granodiorite	6.52	0.0297	0.591	62.8 ± 1.9
SLO 71*	20-28S-13E	Adamellite	6.44	0.0255	0.567	54.5 ± 2.5
SLO 329†	10-29S-13E	Granodiorite	6.48	0.0342	0.901	72.3 ± 2.2

* Analyses by Geochron Laboratories, Inc., Cambridge, Massachusetts, for California Division of Mines and Geology, May 1966.
† Analysis by M. L. Silberman and J. Schlocker, U.S. Geological Survey, Menlo Park, Calif. (M. L. Silberman, 1972, written communication).
[1] See geologic map for precise locations.
[2] Ages calculated from following constants: $\lambda\beta = 4.72 \times 10^{-10}$/year; $\lambda e = 0.585 \times 10^{-10}$/year; $K^{40}/K = 1.22 \times 10^{-4}$ atom percent (in U.S.G.S. sample $K^{40}/K = 1.19 \times 10^{-4}$).

ages determined for biotite from three samples in the mapped area show anomalously young dates of 72.3, 62.8, and 54.5 m.y. (table 3). A single fission track age of 83.3 m.y. was determined on four grains of sphene from La Panza granitic rocks (C.W. Naeser, 1970, personal communication). Other radiometric age-dates for granitic rocks elsewhere in the southern Coast Ranges range from 69 to 117 million years. Based on the oldest date, a rubidium-strontium determination, Evernden and Kistler (1970) consider the granitic rocks of the Coast Ranges (including those the La Panza Range) to have been emplaced during the Huntington Lake epoch, 104-117 m.y. ago. This latter period fits well with the first definite influx of acid plutonic debris of the Atascadero Formation (see above; also see Hart, 1971). However, the younger age-dates cannot be dismissed entirely as being anomalous and may, in part, reflect a later or prolonged period of igneous intrusion.

Unnamed Sandstone and Conglomerate

A thick sequence of Upper Cretaceous sedimentary beds is exposed east of the Rinconada fault in the southeast part of the mapped area. These rocks formerly were mapped as the Vaqueros Formation by Fairbanks (1904). They are somewhat similar to the Atascadero Formation (see above), but on the whole are much coarser grained. This unnamed sandstone and conglomerate sequence constitutes the basal and most westerly part of a very thick sequence of Upper Cretaceous to upper Eocene rocks exposed along the south flank of the La Panza Range (Dibblee, 1968; Vedder and Brown, 1968; Chipping, 1972). In the mapped area, the sequence largely consists of thick beds of coarse-grained, biotitic sandstone and conglomerate. Interbedded are finer grained, more thinly bedded sequences of sandstone, siltstone, and mudstone. The coarse- and fine-grained rocks occur both as individual beds and as bedded sequences as much as several hundred feet thick. At least 4000 feet of strata are present, although numerous minor faults and folds prevent accurate measurement of the thickness. Exposures of the massive coarse-grained rocks tend to be bold, but sagebrush limits access to many of these outcrops. The thin-bedded and finer grained rocks are less resistant and are exposed mainly in streams and road cuts.

The Upper Cretaceous beds are depositional on Cretaceous granite. The basal contact appears to be "gradational" between granite wash and decomposed granite in NE 1/4 sec. 31. Along the Salinas River, the contact is relatively sharp below massive cobble conglomerate and sandstone. Locally, the mid-Tertiary Vaqueros Formation must overlie the Cretaceous beds, but the contact between these units is either faulted or concealed.

Coarse-grained and fine-grained subunits, a few inches to several hundred feet thick, can be recognized locally but are not differentiated on the map. The coarse units are typified by medium to very coarse sandstone, conglomeratic sandstone, and sandy conglomerate exposed in massive beds as much as 20 feet thick or more. Interbedded are sequences of sandstone, siltstone, and mudstone with impure limestone lenses and concretions. The latter sequences are laminated to well bedded. They also display such current features as graded bedding, cross-lamination, scour and fill, intraformational clasts, and sole markings. Other than large-scale cross-stratification and channeling, the coarse massive beds show few internal features. Marine fossils are scattered throughout the section and coaly plant debris is abundant. A thin coal seam is present in one place, and red beds were observed nearby.

The sandstones vary greatly in color, grain size, and sorting; but all are arkosic and generally characterized by large flakes of crinkly biotite. Most of the fresh sandstone is light to medium gray, sometimes with green or brown hues. It weathers readily to tan and yellowish to reddish brown. Some beds are nearly white and others reddish. Induration varies from easily friable and porous to hard, calcitic, and nearly impervious. The main constituents are subangular quartz, plagioclase, and K-feldspar with 1-10% biotite and small amounts of quartzite, chert(?), volcanic and metamorphic rocks, muscovite, and heavy minerals (epidote, pyrite, sphene). The grains are well packed, and many are broken or deformed. Calcite is common as cement and as grain replacements.

Conglomerate lenses and beds to 20 or 30 feet thick are widely associated with the coarse sandstone but are subordinate to the latter. Conglomerate clasts range from pebbles to boulders 2 feet or more in diameter, but small to medium cobbles are most abundant. The clasts consist of well-rounded, resistant volcanic rocks, which are mainly delicately colored porphyries. Granitic rocks are common and locally predominate near the base. These are subangular to rounded. Small amounts of quartzite and other metamorphic rocks, aplite, pegmatite, and vein quartz are generally present as rounded clasts. Sandstone, siltstone, and mudstone are present locally, mainly as intraformational clasts in conglomeratic sandstone.

Mudstone, siltstone, and limestone—some of which are quite sandy—occur as thin beds and laminations in rhythmically bedded sequences with fine- to medium-grained sandstone. They are mostly gray with olive to brownish hues. Fossil debris, which is abundant but not always diagnostic includes foraminifers, Inoceramus sp. (including "prisms"), fish teeth, plant leaves and stems, and palynomorphs.

The unnamed sandstone-conglomerate unit is mainly of marine origin, but it apparently contains some nonmarine beds (coal, red beds). Large-scale cross stratification and channeling reflect strong currents that may be fluviatile in part. Turbidite sequences, as much as 500 feet thick immediately southwest of the Salinas Dam, indicate relatively deep basinal deposition for part of the unit. Foraminifers in mudstone near the base of the sequence in Rinconada Creek probably were deposited in water 200-600 feet deep (C.C. Church, 1966, written communication). Conditions of deposition for most of the coarse massive sandstone-conglomerate sequences may be shallow marine. It is suggested that deposition occurred near shore by rapid development of a prograding sedimentary wedge under varied and changing conditions. A nearby granitic source supplied most of the sediment, although the rounded volcanic clasts indicate a more distant source. Paleocurrent directions for the Upper Cretaceous-lower Tertiary sequence of the La Panza Range are reported to be toward the south (Chipping, 1972).

The age of the unit is latest Cretaceous based on the presence of fossils. Foraminifers from the lower part of the unit (SLO 111 to 113 and 124, table 4) belong to the D-2 zone of Goudkoff, which represents a late Campanian to early Maestrichtian age. Palynomorphs (SLO 123) and *Inoceramus* sp. "prisms" also indicate a Late Cretaceous age for beds perhaps 2000-3000 feet above the base. Diagnostic fossils were not found higher in the sequence in the mapped area.

The unnamed sandstone-conglomerate unit is the most westerly and basal part of a thick Upper Cretaceous to Eocene sequence exposed along the south flank of the La Panza Range (Dibblee, 1968; Vedder and Brown 1968; Chipping, 1972). The unit is lithologically similar to and partly correlative in age with unnamed Upper Cretaceous-lower Tertiary beds mapped by Durham (1968a) in the Santa Lucia Range about 30 to 40 miles to the northwest. The two sequences formerly may have been coextensive and were later separated by large-scale right-lateral offsets along the Rinconada and related faults (see under Structural Features). Other nearby units of partly similar age include the Atascadero Formation (see above), Asuncion Formation (Taliaferro, 1944), and Carrie Creek Formation (Hall and Corbato' 1967).

Table 4. *Upper Cretaceous microfossil localities east of the Rinconada fault.**

	Fossil localities*				
	SLO 111	SLO 112	SLO 113	SLO 124	SLO 123
FORAMINIFERS					
Bathysiphon alexanderi Cushman		×			
perampla Cushman and Goudkoff		×			
sp.	×		×		
Bolivina incrassata Reuss		×			
Bulimina spinata Cushman and Campbell	×	cf	×	×	
Cribrostomoides cretacea Cushman and Goudkoff		×			
Dentalina basiplanata Cushman		×			
sp.				×	
Gaudryina sp.	×		×		
Glandulina manifesta Reuss	×	×			
Gyroidina sp.	×		×		
Lenticulina williamsoni	×		×		
sp.				×	
Marginulina cf *M. elongata* d'Orbigny		×			
Nodosaria cf. *N. alternata*	×		×		
cf. *N. ewaldi* Reuss		×			
spinifera Cushman and Campbell	×	×	×	×	
Silicosigmoilina californica Cushman and Church	×	×	×		
Spiroplectammina anceps Reuss		×			
PALYNOMORPHS					
Dinogymnium sp.					×
Proteacidites thalmanni					×
Other angiosperm pollen					×
MOLLUSKS					
?Inoceramus sp.					×
Identified age	Upper Cretaceous (D-2 zone of Goudkoff)				Upper Upper Cretaceous

* Fossils identified by C. C. Church (SLO 111, 113, 124), R. L. Pierce (SLO 112), and W. R. Evitt (SLO 123).

CENOZOIC ROCKS
Unnamed Conglomerate

A poorly bedded sequence of massive conglomerate with interbedded sandstone constitutes a lenticular or wedge-shaped unit in the northwest part of the map area. The unit is relatively soft and poorly exposed, being concealed by soil and vegetation. Fairbanks (1904) included the conglomerate with his Monterey Formation and Page (1972, figure 3) identified the unit as "Sespe(?) Formation."

The conglomerate is generally thick bedded and poorly sorted. It consists of subangular to subrounded pebble- to boulder-sized clasts as much as 6 feet across. Most of the larger clasts are feldspathic biotitic sandstone typical of the Atascadero Formation. Other large Atascadero-like clasts include conglomerate, pebbly sandstone, fine sandstone, and mudstone. The pebbles and small cobbles probably predominate over the larger clasts. Much of the pebble and small cobble fraction consists of volcanic porphyry and other resistant clasts presumably reworked from the Atascadero unit. A few small clasts of graywacke, mafic volcanic rock, and chert may be from the Franciscan melange.

Feldspathic sandstone and pebbly sandstone constitute interbeds, lenses, and conglomerate matrix in the unnamed conglomerate unit. The sandstone is massive to indistinctly stratified or cross-stratified. It is mostly friable, but locally it is cemented with calcite.

The maximum thickness of the unnamed unit is about 1000 feet at the western boundary of the study area. It thins to the east and locally pinches out to the south. The unit is unconformably depositional on the Atascadero Formation. Its contact with the Franciscan melange is poorly exposed and may either be depositional or faulted. Basal marine sandstone of the overlying Monterey Formation is conformable and locally gradational. In NE 1/4 sec. 25 (proj.), T. 28 S., R. 11 E., the unnamed conglomerate may interfinger with the basal Monterey.

The poorly sorted conglomerate with incompletely rounded boulders of friable sandstone suggests that part of the unit represents debris-flow deposits. The lack of marine fossils and marine sedimentary features, as well as faintly stratified, cross-bedded(?) lenses of coarse pebbly sandstone, indicate a probable fluvial origin for much of the rest of the unit. An abundance of Atascadero-like sandstone and conglomerate detritus, the incomplete rounding of large clasts, and thinning of the unit southward suggests a local source area immediately to the north.

No diagnostic fossils were found, although a single leaf print was noted, in the unnamed conglomerate. The unit's conformable and gradational relationship with the basal Monterey indicates a probable age of late Oligocene or early Miocene. Precise correlations are not possible, but other units of generally similar age and character in the southern Coast Ranges include the basal conglomerate of the Vaqueros Formation (Tvc) and the Simmler Formation(?) (Ts) of the mapped area east of the Rinconada fault and the Berry Conglomerate of the northern Santa Lucia Range (Thorup, 1943; Bramlette and Daviess, 1944; also see lower member of Vaqueros Formation of Durham, 1963).

Simmler Formation(?)

Poorly stratified nonmarine conglomerate beds, composed almost entirely of granitic debris, are exposed east of the Rinconada fault in the vicinity of sec. 9, T. 29 S., R. 13 E. This Oligocene(?) to early Miocene unit is similar in character and stratigraphic position to the Simmler Formation exposed in the La Panza Range a few miles east of the mapped area (Dibblee, 1968). The type locality of the Simmler is located 60 miles to the southeast in the Caliente Range (Hill and others, 1958). In the mapped area, the Simmler Formation(?) is a weakly indurated, easily erodable unit that is highly dissected. It is poorly exposed and densely covered with sagebrush.

The Simmler Formation(?) consists mainly of poorly defined, grayish- to yellowish-brown, massive conglomerate beds of poorly sorted gravel, sand, silt, and clay. The larger clasts are angular to rounded and consist of partly decomposed porphyritic granitic rocks and subordinate pegmatite-aplite that are identical to the granitic rocks exposed to the east. Most of the clasts are cobbles and small boulders, although giant clasts to 15 feet long are present. These are concentrated at the surface as lag. The giant clasts lend the unit a unique character. Weakly consolidated to locally cemented, massive to laminated sandstone, siltstone and claystone beds, partly interbedded with conglomerate, are found in the lower part of the

Simmler(?). The sandstone is moderately to poorly sorted arkose or "granite wash" of alluvial origin. The clay beds are probably lacustrine.

The maximum thickness of the Simmler Formation(?) is uncertain because exposures are few and bedding is indistinct, but it is inferred to be about 600 feet. The finer clastics appear to be confined to the lowest 100-200 feet. The Simmler(?) wedges out gradually to the northwest and abruptly to the southeast. It may interfinger with the marine Vaqueros Formation to the southwest as indicated by the outcrops in E 1/2 sec. 16, T. 29 S., R. 13 E. Elsewhere, the Simmler(?) grades upward, locally by interbeds, into coarse arkosic sandstone and conglomeratic sandstone of the conformably overlying marine sandstone of the Vaqueros. It unconformably overlies the Upper Cretaceous granitic rocks.

A lack of fossils in the Simmler Formation(?) prevents the assignment of a precise age. Gradational relations with the overlying and partly interfingering(?) Vaqueros, which contains fossils of early Saucesian age, indicate an age of late Oligocene to early Miocene.

Much of the Simmler(?) is unsorted and massive and most likely represents ancient debris or mudflows derived from weathered granitic rocks of the La Panza Range. The lower part of the unit mainly suggests fluvial and minor ponded conditions.

In the mapped area, the Simmler Formation(?) is correlative in stratigraphic position and approximate age to: 1) the basal nonmarine conglomerate member of the Vaqueros Formation 4 miles to the southeast and 2) the unnamed nonmarine conglomerate 8 miles to the west-northwest, near Paradise Valley. However, both of these units are somewhat different in composition. The Simmler(?) also is lithologically identical to the giant-boulder conglomerate of the Tierra Redonda Formation (early to middle Miocene) of Durham (1968a) which is exposed near Harris Valley, 33 miles to the northwest. These outcrops were examined briefly in May 1967. However, the precise age of that conglomerate or its stratigraphic position within the Tierra Redonda is uncertain (Durham, 1971, personal communication). It seems likely that the conglomerate unit of Harris Valley formerly lay southwest of and was contiguous with the Simmler(?) of the mapped area.

Vaqueros Formation

The Vaqueros Formation is used herein to designate the predominantly coarse-grained marine sedimentary rocks that underlie the Monterey Formation. It essentially includes the same rocks mapped as Vaqueros by Fairbanks (1904) but excludes the thick sequence of unnamed Upper Cretaceous beds east of the Rinconada fault in the southeast part of the mapped area. The Vaqueros was originally named by Hamlin (1904) for sandstone beds exposed in Vaqueros Creek to the northwest. The unit was defined and restricted by Thorup

(1944) and later redefined to include the nonmarine (?) Berry Conglomerate (lower Vaqueros) by Durham (1963). An unnamed nonmarine(?) conglomerate member, which may be a lateral equivalent of the Berry, is differentiated locally by this writer (Tvc on map). The term Vaqueros has been used throughout the southern Coast Ranges as a formation and over a still-greater region as a megafaunal stage ("Vaqueros Stage" of Weaver and others, 1944). In spite of its confusing usage, the Vaqueros Formation (unrestricted) is reasonably well suited to this writer's needs.

In the mapped area, the Vaqueros Formation is exposed on both sides of the Rinconada fault, although the unit has contrasting characteristics and thicknesses on opposite sides of the fault. West of the fault, the Vaqueros is a marine sandstone that is 125 feet thick or less and locally is absent. East of the fault, it varies in thickness from 20 to 700 feet or more and is of varied lithologic character (marine sandstone, siltstone, mudstone, and nonmarine(?) conglomerate) along the strike. The Vaqueros units west and east of the Rinconada fault are described separately below.

West of the Rinconada Fault

The Vaqueros Formation is exposed intermittently along the northeast flank of the Santa Lucia Range where it rests unconformably on the Atascadero Formation. It consists mainly of 5- to 10-foot thick beds of massive, gritty, arkosic sandstone with sandy and pebbly reefoid limestone, pebble conglomerate, and pebbly sandstone that constitute a sequence 10 to 30 feet thick near Atascadero and up to 125 feet thick a mile southeast of Trout Creek. The Vaqueros is conformably overlain by and locally is gradational with the Monterey Formation. Elsewhere the Vaqueros sandstone is either too thin to map (and is included with the basal Monterey) or is absent.

The sandstone is mostly coarse grained, pebbly, hard, poorly to moderately sorted, and light to medium gray brown. The grains are mainly subangular to subrounded quartz, plagioclase, K-feldspar, rock fragments (volcanic rocks, chert, clastic sedimentary rocks), and shell debris that are loosely packed. Calcite cement is abundant and results in the sandstone's being resistant and hard; where absent, the sandstone is very porous and friable. The sandstone grades into or is interspersed with lenses of pebble conglomerate on one hand and sandy calcarenite and bioclastic limestone on the other. Recognizable shell debris—mollusks, calcareous algae, foraminifers—that is locally phosphatic shows the unit to be of shallow marine origin. The terrigenous grains indicate a nearby source of granite (or reworked coarse arkosic sandstone), Franciscan melange, and sedimentary rocks.

Fossils which are heavy-shelled shallow marine forms, are generally fragmented and were not collected for identification. However, incomplete specimens of large pectinids, similar to those found elsewhere in the Vaqueros, were noted. According to Loel and Corey (1932), fossils (Pecten magnolia,

P. vanvlecki, etc.) found in the Vaqueros Formation north of Atascadero Creek are meager but indicate an upper "Vaqueros age" or "Oligo-Miocene" age of Weaver and others (1944). Early Saucesian Foraminifera from the immediately overlying Monterey beds (table 5, localities 168 and 205) in the Atascadero quadrangle also indicate the Vaqueros to be late Oligocene and/or early Miocene. A younger age is suggested to the southeast, near Trout Creek, in the Lopez Mountain quadrangle, where the oldest Monterey fossils sampled (SLO 119, 133, 134) are of a Relizian ("early middle" Miocene) age. However, this area is structurally complex, and lower beds of the Monterey may have been faulted-out locally and the fossil data may be inadequate.

Although the Vaqueros Formation is of similar character and thickness where developed west of the Rinconada fault, it apparently transgresses time upward from north to south (figure 6). In a gross sense, it may be viewed as a basal transgressive sand of the Monterey Formation. Apparent age differences and the presence of locally derived sediment in the Vaqueros and the locally underlying unnamed nonmarine conglomerate suggest local tectonic changes during late Oligocene to middle Miocene time.

East of the Rinconada Fault

The Vaqueros Formation east of the Rinconada fault is of variable age, composition, and thickness. From north to south, the unit changes from a resistant gritty, calcareous sandstone facies of rather constant thickness (20-100 feet) north and east of the Salinas River (northwest Santa Margarita quadrangle) to a thicker (200-700+ feet) and more varied facies of softer sandstone, siltstone, mudstone, and conglomerate south and west of the Salinas River (south Santa Margarita and northeast Lopez Mountain quadrangles). These will be referred to as the northern and southern facies. The latter also includes a thick nonmarine(?) conglomerate member (Tvc on geologic map).

The northern facies extends from Rocky Canyon on the north to a point about half a mile south of where the Vaqueros is crossed by the Salinas River. This segment of the Vaqueros Formation mainly consists of hard, massive, thick-bedded, brownish-gray, calcareous sandstone and pebbly sandstone that grades downward into a thin basal conglomerate unit. These rocks, which weather reddish or yellowish brown, rest on weathered granitic rocks. Thin lenses of sedimentary breccia, composed of poorly sorted, angular granitic debris, commonly separate the marine Vaqueros from the granite. Beds of hard, brown, light-gray weathering dolomite and sandy dolomite are present locally in the upper part of the Vaqueros where it grades, generally rapidly, into the overlying dolomite and shale of the Monterey Formation. Excluding the basal sedimentary breccia, the Vaqueros averages about 50 feet in thickness, ranging from 20 to 100 feet. The underlying sedimentary breccia generally is thinner or absent, although in one place where it is about 200 feet thick it is tentatively mapped as Simmler Formation(?).

Sandstone of the northern facies is very coarse to pebbly and mostly composed of subangular grains of quartz, plagioclase, K-feldspar, granitic rocks, and some fossil debris, with only minor amounts of volcanic rocks, quartzite, and other grains. The grains are moderately to poorly sorted, loosely packed, and set in a micritic to sparry calcite cement. Locally present is a nearly white, massive, well-sorted, medium-grained, arkosic sandstone. The white sandstone is most prominent south of the Salinas River where the northern facies gives way to the southern facies. The lower Monterey of this area also reflects a lateral change, being sandy, silty, and less dolomitic near its gradational basal contact.

The southern facies of the Vaqueros Formation is a poorly stratified sequence of sandstone, conglomerate, siltstone, mudstone, and limestone. The sequence is poorly exposed and structurally disturbed. It is typified by massive, nearly white to pale orange-gray, medium-grained, well-sorted, arkosic sandstone in beds as much as 15 to 20 feet thick. Much of the rest of the sandstone is buff to light gray and, although of similar composition, is coarse, moderately to poorly sorted, and commonly conglomeratic. Some also is essentially uncemented and friable. The sandstone is silicified locally close to the Rinconada fault.

Conglomerate lenses and beds interfinger with the coarse sandstone, particularly in the lower part of the unit. The clasts are rounded to angular pebbles, cobbles, and small boulders of granitic rocks. An exception is the unnamed conglomerate member (see below) that contains abundant clasts derived from adjacent unnamed Upper Cretaceous beds. The conglomerate varies from hard and calcareous to friable and uncemented.

Soft siltstone, fine-grained sandstone, mudstone, and sandy claystone of buff to brownish gray are locally abundant and apparently interbedded with the white sandstone. Some of these fine-grained rocks are calcareous, foraminiferal, and bituminous. Local lenses of limestone (calcarenite) and sandy limestone are interspersed and gradational with the sandstone. Beds of siliceous sandstone, siltstone, tuff(?), and porcelanite also are exposed along the Rinconada fault zone. These may be fault slivers of the Monterey Formation but most likely are a Monterey facies interbedded with the upper part of the Vaqueros.

Unnamed Conglomerate Member

A nonmarine(?) conglomerate and conglomeratic sandstone unit, mapped as Tvc, is present in the vicinity of Rinconada Creek and Las Pilitas Road. The rocks are poorly stratified and massive to coarsely cross-bedded. Cut and fill features are common. The conglomerate clasts mainly are subrounded pebbles and cobbles, although small boulders are present in places. Clasts are mostly volcanic porphyry with moderate amounts of granitic rocks and a small percentage of quartzite, dark metamorphic or volcanic rocks, biotitic sandstone, and mudstone. The sandstone is

a very coarse, moderately well-sorted, friable arkose. Some biotite is present but not the crinkly kind characteristic of the Upper Cretaceous sandstones. The unnamed conglomerate unit consists of debris derived from the granitic and unnamed Upper Cretaceous rocks, which are exposed to the north and east, respectively.

The basal contact of the unnamed conglomerate member is not exposed, but muddy brown sandstone near the base of the unit suggests that the member is locally depositional on the Upper Cretaceous unit. The conglomerate unit is in fault contact with the Upper Cretaceous beds to the east and apparently grades northwestward into marine sandstone of the Vaqueros. Pebbly interbeds in the lower 50 feet of the white Vaqueros sandstone beds south of Las Pilitas Road suggest a gradational contact between the marine Vaqueros and its nonmarine(?) conglomerate member. A maximum thickness of perhaps 500 feet is estimated for this member. It is unfossiliferous and its age is uncertain. Based on lithology and apparent stratigraphic relations with the marine Vaqueros, the conglomerate sequence is probably late Oligocene or early Miocene in age.

Thickness and Conditions of Deposition

As indicated above, the Vaqueros Formation east of the Rinconada fault varies from 20 to 100 feet thick in the north. The southern facies, including the unnamed conglomerate member, varies in thickness from about 200 feet at its northern end to a maximum of at least 700 feet at its southern end. The thick conglomerate member and possibly some of the thin basal conglomerate lenses northwest of it may be of nonmarine (fluvial) origin. The rest of the unit is marine, the sandstone and limestone representing shallow, nearshore facies and the siltstone-mudstone somewhat deeper water. Middle to lower bathyal depths (about 1500-6000 feet) are indicated by foraminiferal faunas in samples SLO 78, 79, and 80 (table 5) according to R.L. Pierce (1966, written communication). The terrigenous clastic sediment presumably was derived from rocks of the ancient La Panza Range that lay to the east. Local tuffaceous(?) beds suggest nearby volcanism.

Age and Correlation

Diagnostic fossils in the Vaqueros Formation were found only in the southern facies. Foraminifera from four localities (SLO 78, 79, 80, and 327) northeast of Santa Margarita indicate an early Saucesian to early Relizian(?) age for much of the Vaqueros. Fossils from the adjacent and overlying(?) Monterey Formation (SLO 97 and 115) indicate that the Vaqueros is locally no younger than early Relizian. To the southeast of State Route 58, however, foraminifers from the two localities sampled (SLO 105 and 108) show that part of Vaqueros is of Relizian age. The age of most of the unit in this area remains undetermined. The thin northern facies of the Vaqueros is directly overlain by basal Monterey beds of early Saucesian age in Rocky Canyon (SLO 225) and of early Relizian age 2.5 miles south of there (P.B. Smith, 1971, personal communication),

Table 5. Fossil foraminifers from the Monterey and Vaqueros Formations of the Santa Margarita area, San Luis Obispo County, California. [Abbreviations: X, present; cf, similar form present; var., variety of form identified; ?, questionable identification; sp., species not determinable.]

	Fossil localities (SLO numbers)																													
	Within and east of Rinconada fault zone												West of Rinconada fault zone																	
	47	78	79	80	97	101	105	108	115	130	225	327	116	119	132	133	134	140	154	155	156	157	158	160	161a	167	168?	169?	205?
Angulogerina occidentalis (Cushman)																											cf		
Anomalina glabrata Cushman			X		X			X														X							X
salinasensis Kleinpell							cf																						
sp.	X																												
Baggina californica Cushman	X																									X			X
cancriformis Kleinpell																													X
robusta Kleinpell													X	X	X	X	X	X	X	X	X			X	X	X	X		X
Bolivina advena Cushman	X		X	X	X		X		X														X	X	X				X
advena striatella Cushman							X	X			X		X	X		X	X		X					X				X	X
californica Cushman							X		X				X	X				X						X					
conica Cushman																													
cuneiformis Kleinpell									X																				
floridana Cushman																													
guadeloupae Parker						X		X	X	X	X		X							X					X		X		X
imbricata Cushman																													
imbricata Cushman var.												cf						cf		cf					cf				
marginata Cushman																													
adelaidana Cushman and Kleinpell																													
gracillima Cushman and Laiming																											X		X
pisciformis Galloway and Morrey																											X		
tumida Cushman																													
Buccella frigida (Cushman)															X	X	X	X	X		X	X		X	X		X		X
Bulimina inflata Seguenza																													
inflata alligata Cushman and Laiming		X																											
montereyana Kleinpell				cf																	X								
ovula d'Orbigny																													
cf. B. seigerinaeformis Cushman and Kleinpell																													
subfusiformis Cushman	X		X		X	X	X	X	X	X		X	X	X	X	X	X	X	X		X	X		X	X		X	X	X
Buliminella curta Cushman																													
elegantissima (d'Orbigny)																					X								
subfusiformis Cushman and Kleinpell																													
Cancris baggi Cushman and Kleinpell																				X									
sagrai (d'Orbigny)																											X		X

continued on page following

Table 5. Fossil foraminifers from the Monterey and Vaqueros Formations of the Santa Margarita area, San Luis Obispo County, California. [Abbreviations: X, present; cf, similar form present; var., variety of form identified; ?, questionable identification; sp., species not determinable.] -(continued)

Fossil localities (SLO numbers)

	Within and east of Rinconada fault zone												West of Rinconada fault zone																
	47	78	79	80	97	101	105	108	115	130	225	327	116	119	132	133	134	140	154	155	156	157	158	160	161a	167	168²	169	205³
Cassidulina laevigata carinata Cushman		X																											
cf. C. limbata Cushman and Hughes																													
margarita Karrer		X																											
sp.																													
Cibicides americanus Cushman			X			X																				X			
americanus crassiseptus Cushman and Laiming		X	X																										
floridanus (Cushman)		X																											
pseudoungerianus evolutus Cushman and Hobson																													
relizensis Kleinpell																										X	X		
sp.																								X					
Dentalina obliqua Linné					X						X																		X
quadrulata Cushman and Laiming										X														X		X			
sp.																													
Epistominella gyroidinaformis (Cushman and Goudkoff)								X																				X	
relizensis Kleinpell																			X	X		X	X	X	X				
subperuviana (Cushman)																			X	X	X	X	X	X	X				
Eponides umbonatus (Reuss)											cf																		
Globigerina bulloides d'Orbigny										X									X				X	X	X				
spp.						cf																			cf				
Lenticulina hughesi Kleinpell		X																											
laimingi Bandy and Arnal							cf																		cf				
miocenica (Chapman)																X	X	X	X	X		X	X	X					
simplex (d'Orbigny)			X												X	X	X	X						X	X	X	X		
smileyi (Kleinpell)																				X		X	X	X	X				
sp.																													
Nodogenerina advena Cushman and Laiming		X	X			X	X		X	X	X	X						X	X			X		X	X	X		X	
Nonion costiferum (Cushman)		X	X								X	X							X	X	X	X	X	X		X		X	
incisum (Cushman)												X																	
Nonionella miocenica stella Cushman										X														X					
sp.																													
Planulina appressa Kleinpell																									X				
baggi Kleinpell														X															X
dubia (Egger)										X																			?

…continued on page following.

Table 5. Fossil foraminifers from the Monterey and Vaqueros Formations of the Santa Margarita area, San Luis Obispo County, California. [Abbreviations: X, present; cf, similar form present; var., variety of form identified; ?, questionable identification; sp., species not determinable.] -(continued)

Fossil localities (SLO numbers)

Species	Within and east of Rinconada fault zone												West of Rinconada fault zone																
	47	78	79	80	97	101	105	108	115	130	225	327	116	119	132	133	134	140	154	155	156	157	158	160	161a	167	168	169	205
Plectofrondicularia californica Cushman and Stewart		X	X																							cf	X		X
miocenica Cushman																													X
directa Cushman and Laiming	X																												
Pullenia miocenica Kleinpell	X																							X					
miocenica globula Kleinpell																													
multilobata Chapman																													
Saracenaria beali (Cushman)	X	X																									X		
dubia Neugeboren																													
Siphogenerina branneri (Bagg)		X						cf																X	X		X		
collomi Cushman															X	X	X				X			X	X				X
hughesi Cushman							X		X				X	X	X	X				X				X					X
kleinpelli Cushman											X					X	X								X				
cf. *S. tenua* Cushman and Kleinpell																	?												
transversa Cushman		X	X		X		X	X	X	X	X	X	X	X	X	X	X	X		X	X			X	X	X	X		X
sp.																											?		
Uvigerina cf. *U. auberiana* d'Orbigny																											X		
Uvigerinella californica Cushman								X								X				X				X					X
californica parva Kleinpell				X					X																				
obesa Cushman									X																				
impolita Cushman and Laiming																													
sp.		X	X			X			X				X	X	X	X	X	X		X				X	X				
Valvulineria californica Cushman																X	X							X	X		X		
californica var. obesa Cushman																X	X			X					X				
cf *californiensis* Cushman and Laiming																													
depressa Cushman		X		X	X		X	X			X		X	X	X	X	X	X		X	X			X	X				
williami Kleinpell			cf																										
sp.						X																							
Virgulina bramlettei Galloway and Morrey		X				X	X						X										X						
californiensis Cushman																					X								
californiensis grandis Cushman and Kleinpell																													
delmontensis Cushman and Galliher				X			X X												X X				X						
Indicated age (and paleontologist)[1]	Luisian (C)	Lower Saucesian (P)	Relizian or Saucesian (P)	Saucesian? (P)	Upper Relizian (C)	Miocene? Relizian? (C)	Relizian or Saucesian (P)	Upper Relizian (C)	Lower Relizian (P)	Upper Relizian (C)	Lower Saucesian (C)	Saucesian (C)	Upper Luisian	Upper Relizian (S)	Luisian (S)	Lower Relizian (S)	Lower Relizian (S)	Upper Relizian (C)	Relizian or lower Luisian (S)	Upper Relizian (C)	Luisian (S)	Middle Miocene (S)	Relizian? (C)	Upper Relizian (C)	Lower Relizian (C)	Saucesian (C)	Saucesian (S)	Lower Mohnian (C)	Upper Saucesian (S)

[1] C = C. C. Church, Consultant, Bakersfield.
P = R. L. Pierce, U.S. Geological Survey, Menlo Park.
S = P. B. Smith, U.S. Geological Survey, Menlo Park.

[2] Same as U.S. Geological Survey locality: SLO 168 = Mf1201; SLO 205 = Mf1500.

which provides minimum ages for the Vaqueros. Based on limited microfossil data, it would appear that the Vaqueros Formation varies somewhat in age from place to place, reflecting locally varied and changing conditions during late Oligocene to middle Miocene time. The southern facies appears to be partly a lateral sandy facies of the lower Monterey and partly a lateral equivalent of the northern facies of the Vaqueros (figure 6).

In the mapped area, the Vaqueros east of the Rinconada fault correlates with the Vaqueros and lower Monterey west of that fault. It is believed, however, that the Vaqueros units on opposite sides of the fault were initially deposited in separate areas that are now juxtaposed as a result of large-scale right-lateral displacement along the Rinconada fault. Correlations outside the mapped area can be made with many units of the Santa Lucia Range. To the northwest, it is equivalent in age, in part or whole, to the Vaqueros Formation, Berry Conglomerate, Tierra Redonda Formation, and Sandholdt Member of the Monterey Formation (e.g.,

Thorup, 1944; Durham, 1963, 1968a, 1968b). Specifically, the Tierra Redonda Formation of Durham (1968a) is similar in age and lithology and is believed to be a former southwesterly extension of the southern facies of the Vaqueros Formation of the mapped area. To the south and southeast, rock units of equivalent age include the Rincon, Obispo, and Point Sal Formations and possibly the Vaqueros Formation (Hall and Corbato', 1967; Woodring and Bramlette, 1950). Equivalent-aged rocks to the east, along the northeast flank of the La Panza Range, are the Vaqueros, lower Monterey and possibly the Simmler Formation (Dibblee, 1968).

Monterey Formation

The Monterey Formation is a well-bedded marine Miocene sequence composed mainly of siliceous and calcareous sedimentary rocks. It essentially includes those rocks formerly designated as Monterey Shale by Fairbanks (1904). The Monterey and its informal subdivisions are similar in age and lithology

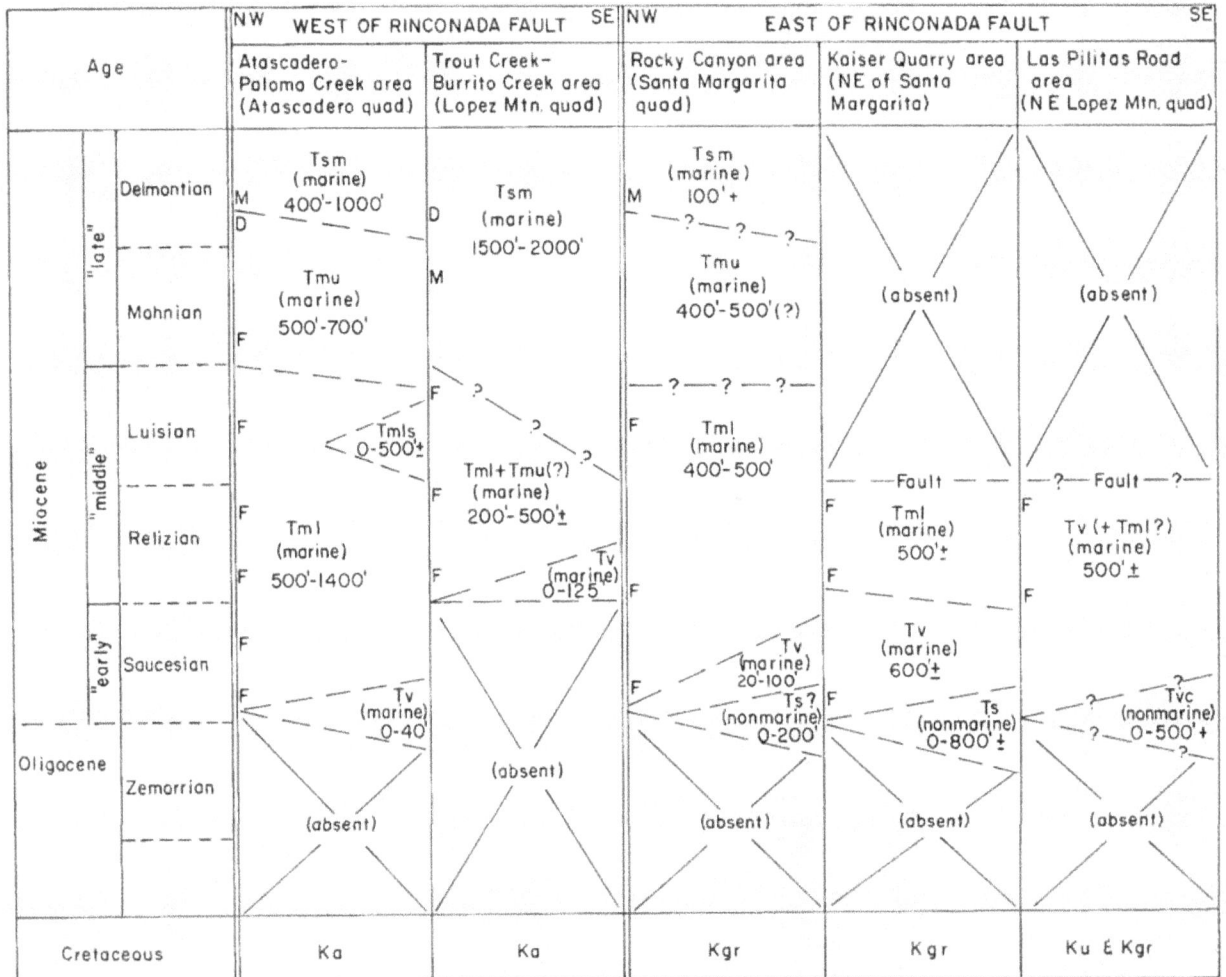

Figure 6. Correlation chart of middle Tertiary rock units, Santa Margarita area. Shows thickness and age variations and fossil control (F = Foraminifera; D = diatoms; M = megafossils). See plate 1 for unit symbols.

and are lateral extensions of the Monterey Shale units exposed to the northwest and north (Durham, 1963, 1968a, 1968b; Smith and Durham, 1968; Smith, 1968; Burch and Durham, 1970). The type area is near the city of Monterey 100 miles to the northwest. Bramlette (1946) studied and described the siliceous rocks of the Monterey in detail in the Los Angeles to San Francisco region. Because the unit contains diverse rock types and the term "shale" is somewhat of a misnomer, the designation of Monterey Formation is used herein.

In the study area, the Monterey Formation is exposed in three places: 1) a central synclinal belt (Santa Margarita syncline) between the Rinconada and "Nacimiento" fault zones: 2) east of the Rinconada fault along the faulted western margin of the La Panza Range: and 3) a small western area within the "Nacimiento" fault zone near the western margin of the area. Upper siliceous (Tmu) and lower calcareous (Tml) members can be recognized in the central and eastern areas. Only the lower member and associated mafic volcanic rocks (Tmlv) are exposed in the western area. Because of certain differences in the lithology, age, and probable large-scale late Tertiary movements along the Rinconada fault, and possibly the "Nacimiento" as well—these areas are discussed separately below.

Central Synclinal Belt

This area is discussed first because it contains the best exposures and most complete sequences of the Monterey Formation, its subdivisions, and its facies changes. The formation can be traced for at least 17 miles, in the surface and subsurface, along the faulted and folded northwest-trending Santa Margarita syncline. It ranges in thickness from about 200 feet at the southeastern end of the syncline to about 2000 feet near Atascadero. Two distinct members or facies can be recognized: a lower calcareous member and an upper siliceous member. The contact between the members can be mapped readily west of Atascadero. Elsewhere the units can be separated only with difficulty, or not at all, because of structural deformation.

Lower Member

This unnamed member is characterized by well-bedded, foraminiferal and calcareous shale, mudstone, and siltstone. It is commonly altered to or interbedded with siliceous shale, porcelanite, dolomite, and sandstone with minor amounts of algal limestone, conglomerate, bentonite, and pelletal phosphorite. The calcareous and dolomitic rocks are bituminous, giving a petroleum odor upon breaking or on the application of acid.

The lower member of the Monterey Formation conformably overlies the Vaqueros Formation where the latter is thick enough to map. Elsewhere, the Monterey unconformably overlies the Atascadero Formation or Franciscan melange. The upper contact of the lower member west of Atascadero is drawn at the base of the resistant cherty and porcelaneous beds of the upper member. Here, the upper part of the lower member is mainly soft sandstone, locally with a thin sequence of foraminiferal siltstone at the top. Elsewhere, the contact is gradational. The transition zone between the upper and lower members may be 200-300 feet thick in some places. The lower member ranges in thickness from 200 or 300 feet southeast of Trout Creek to about 1400 feet west of Atascadero. A thickness of less than 200 feet is indicated in the subsurface at the Santa Margarita Land and Cattle Co. well, east of Santa Margarita(table 9). Except in roadcuts and creeks, only the more resistant beds are exposed. The unit weathers to a clayey soil (locally sandy) that supports grass and forest vegetation.

The calcareous shale, mudstone, and siltstone, which dominate in the member, form massive to laminated beds generally less than 6 inches thick. They are usually gray brown where fresh but weather to lighter shades of buff, tan, cream, and pale gray. They are composed of various mixtures of clay, silt, and organic matter. Where foraminifers predominate, a punky limestone—often termed "foraminiferite"—is formed. Some of these beds are slightly siliceous to porcelaneous, suggesting diagenetic alteration from diatomaceous (Bramlette, 1946) and/or tuffaceous rocks. Thin beds of gray-brown, soft, plastic bentonite less than 6 inches thick indicate the former presence of vitric tuff. Small phosphate pellets occur as phosphorite layers less than 3 inches thick and also are sparsely disseminated throughout the member.

Light-gray to grayish-orange arkosic sandstone beds as much as 20 feet thick are widely distributed in the lower member of the Monterey. They are most notable, however, in the basal part of the unit and as a thick sandy facies north of Atascadero Creek. The sandstone is fine to coarse and locally pebbly. It varies from hard and cemented (mainly by calcite, locally dolomite, and possibly silica) to soft and uncemented. Most of it is moderately well sorted, but some is poorly sorted and looks "dirty". It contains mainly quartz and feldspars and the basal sandstone contains abundant lithic grains (volcanic rock, chert, sandstone, siltstone) of local derivation. Biotite is not abundant. Mollusk shells and shell debris are common in the basal sandstone. A tongue of sandstone at the top of the member is partly differentiated west of Atascadero (Tmls on plate 1). This unit, which may be several hundred feet thick, pinches out to the south and interfingers with shale, siltstone, and sandstone to the north. The sandstone is mostly a soft, massive, well-sorted arkose. Sandstone dikes 3 inches to 2 feet thick were noted below the sandstone units and along the east side of Graves Creek. These are similar in composition to nearby sandstone interbeds. Clastic dikes of the latter area were described by Newsom (1903).

Light-gray to cream-colored, dense dolomite also is common within the lower Monterey. It occurs throughout the member as massive to laminated, resistant beds a few inches to 5 or 10 feet thick. Dolomite, especially in association with sandstone, helps to characterize the lower 100-200 feet of the Monterey near Trout Creek. The dolomite lacks fossils and is believed to be a diagenetic alteration product derived from the calcareous rocks.

Lenses and beds of hard impure limestone are present locally near the base of the Monterey Formation. These are probably reefoid, being composed of algae, mollusks, foraminifers, and their debris. Impurities include sand, pebbles, and phosphatic pellets and shell fragments.

Upper Member

This unit is characterized by siliceous, often rhythmically bedded rocks that weather to very light gray or tan. They form an interbedded sequence of porcelanite, opaline chert, mudstone, shale, diatomite, siltstone, tuff, and varieties of these. Scattered beds of dolomite, pelletal phosphorite, and bentonite are also present. The upper member grades, partly by interbeds, through transition zones as thick as 200-300 feet into the conformably overlying Santa Margarita Formation and underlying lower member of the Monterey. The upper Monterey has a maximum thickness of 500-700 feet in the northern part of the mapped area and greatly thins to the southeast by interfingering with the Santa Margarita Formation. Most of the unit is well exposed, partly because it contains resistant beds and partly because it develops only a thin, impoverished soil. It is commonly covered by chaparral but locally is grass-covered or forested.

The porcelanite forms beds less than an inch to about 3 feet thick. It is hard, brittle, breaks with a hackly fracture, and has a dull luster. Subvitreous to waxy chert, often in beds 1-3 feet thick, is less common. These siliceous rocks are composed mostly of cristobalite or opal and probably are altered diatomaceous and tuffaceous rocks. Associated are laminated to thin-bedded shale and mudstone that are brown to greenish gray when fresh but weather to tan colors. Although some of these rocks are soft and clayey and have a conchoidal fracture, most are hard and brittle and have a hackly fracture. Probably all gradations exist between the shale-mudstone and the porcelanite. Rare thin shale beds contain foraminifers, and many of the silicified beds show scattered molds of these microfossils. Some of the semi-siliceous shale is very absorbant and may contain abundant clay. Interbedded are thin beds of pelletal phosphorite set in a porcelaneous matrix. Bentonite, plastic when moist, occurs in sparse beds a few inches thick.

Nearly white diatomite locally is concentrated in the upper 50 feet of the Monterey north of Santa Margarita. The diatomite is composed almost entirely of marine diatoms, sponge spicules, and related siliceous microfossils when pure; but it generally contains various mixtures of ashy tuff, sand, silt, and clay. It is a punky, porous, lightweight rock that is found in thin to thick, laminated to massive beds. It contains concretionary beds of porcelanite and chert and grades laterally into thick sequences of these altered siliceous rocks.

Pale-gray, unaltered, vitric tuff that is soft and porous, occurs as sparse 2-inch- to 1-foot-thick interbeds in both the diatomite and porcelaneous sequences. It contains tiny scattered crystals of plagioclase. In one place, the tuff is saturated with bitumins. The tuff probably is an unaltered counterpart of the bentonite.

Massive, white to light-gray, coarse, locally pebbly, arkosic sandstone is common in the uppermost part of the Monterey Formation. Fine-grained sand also occurs as partings and thin beds in diatomite. Well consolidated beds of muddy sandstone and sandy mudstone are common as interbeds with siliceous and porcelaneous rocks. The sandy beds consist of 30-70 percent sand set in a siliceous mudstone or porcelaneous matrix. The sand grains (quartz and feldspars) are both rounded and angular. Very elongate crystals and broken grains and shard-like ghosts in the matrix, as seen in thin sections, suggest that some of the muddy sandstones are altered tuffaceous rocks. A single bed of sandy vitric tuff (divitrified), noted in one place in ashy diatomite, may be an unaltered (i.e., not silicified) equivalent of the siliceous muddy sandstones.

The compositions of some of the fine-grained rocks and their alteration products were revealed by x-ray diffraction analyses. Two samples of relatively soft mudstone (conchoidal fracture) showed abundant montmorillonite with decreasing amounts of cristobalite, quartz, plagioclase, K-feldspar, and mica. Two samples of hard, brittle shale (hackly fracture), which is interbedded with the mudstone, contained abundant cristobalite and moderate amounts of quartz and plagioclase. The x-ray pattern revealed no ordered clay, although these rocks are highly absorbant. A single sample of subvitreous chert from a concretionary bed in ashy diatomite contained a large amount of cristobalite with minor quartz and plagioclase. A nearby sample of the somewhat impure parent diatomite showed moderate amounts of plagioclase, quartz, and montmorillonite with only a trace of cristobalite. Apparently the diatoms are opaline and, therefore, x-ray amorphous. A little fresh volcanic glass is present, and this also must be amorphous.

East of the Rinconada Fault

The Monterey Formation east of the Rinconada fault is lithologically and stratigraphically similar to the Monterey of the central synclinal belt. Lower calcareous and upper siliceous members can be identified and crudely mapped north of Rocky Canyon where the Monterey is about 1000 feet thick. South of there, however, the upper member is not obviously exposed and may not be present. North and east of the Salinas River, the Monterey conformably overlies a thin Vaqueros sequence. It appears to grade normally upward through a series of diatomaceous, sandy, and tuffaceous beds into sandstone of the Santa Margarita Formation. Although relations are not clearly exposed south and west of the Salinas River, the lower Monterey appears to grade laterally into sandy and silty beds of the Vaqueros Formation.

Because of extensive faulting and folding, the Monterey is poorly exposed. The calcareous rocks tend to be covered with soil and a conspicuous calcareous tufa or caliche debris that shows contorted laminae. Virtually all of the rock types observed west of the Rinconada fault are present east of the fault (see Central Synclinal Belt for descriptions).

Western Area

Rocks typical of the lower Monterey Formation, associated with mafic volcanic rocks, are exposed adjacent to and west of Paradise Valley at the western margin of the map area. The unit forms a westerly trending synclinal sequence that rests conformably on the unnamed conglomerate beds and unconformably on the Atascadero Formation, the whole forming a thrust plate above Franciscan melange. The lower 100-300 feet of the Monterey sequence consists of typical shale, mudstone, dolomite, and sandstone that is commonly calcareous and foraminiferal. A 10-15 foot thick Vaqueros-like sandstone and pebbly sandstone occupies the base of the Monterey. Overlying the basal sedimentary rocks, the upper contact of which is locally baked (discolored and siliceous), are mafic volcanic rocks. These are approximately delineated (Tmlv) on the geologic map but were not studied in detail. An upper sequence of sedimentary rocks of the lower Monterey either interfingers with or overlies the mafic volcanics locally. The mafic volcanic unit has a maximum thickness of about 1000 feet.

Where exposed, the volcanic rocks consist of dark gray-green olivine basalt and diabase that readily weather to brownish rock and soil. The relatively fresh diabase consists of abundant plagioclase laths, intersertal to ophitic clinopyroxene (mainly augite), anhedral olivine, and various alteration products. Probable pillow structures suggest that some of the rocks may be of marine extrusive origin. However, much of this unit could be intrusive. The volcanic rock locally consists of silicified tuff(?) composed of cristobalite, plagioclase, calcite, and unidentified minerals (x-ray amorphous?). This siliceous altered rock is mottled a light blue-gray and partly resembles the zeolitized tuff of the Obispo Formation south of San Luis Obispo (Hall and Surdam, 1967). Rhyolite tuff is associated with a "diabase sill" in the lower part of the Monterey Formation a few miles to the west of the mapped area (Page, 1970b, figure 3).

Origin and Conditions of Deposition

The Monterey Formation is entirely of marine origin as indicated by the abundant marine fossils distributed throughout it. The lowermost and uppermost beds contain coarse clastic sediment and fossils (mollusks, algae) of shallow water, nearshore origin. Most of the unit, however, contains foraminifers representative of bathyal to outer neritic depths. Similar rocks and foraminifers of the Monterey north of the La Panza Range near Indian Creek are considered to represent depositional depths of about 4000 feet (lower calcareous Monterey)

to about 150 feet (upper siliceous Monterey) (Smith, 1968).

The principal primary sediments must have consisted of terrigenous sand, silt and clay, foraminiferal and diatomaceous oozes, tuffaceous sediment, minor reefoid deposits, phosphate pellets, and various admixtures of these. Some of the calcareous sediment and most of the diatomaceous and tuffaceous sediments were subsequently altered by widespread diagenesis to dolomitic, siliceous, and montmorillonitic rocks. Deposition of these multiple-source deposits and subsequent alteration account for the broad spectrum of rock-types in the Monterey. To understand the Monterey in all its aspects, it is useful to view the unit as a continuum of its various primary constituents and their alteration products.

Age and Correlation

The Monterey Formation ranges in age from early to late Miocene. Fossils indicate that the formation varies in age both within and between the central and eastern outcrop areas (figure 6 and tables 5, 6, 7). No diagnostic fossils were identified in the limited exposures of the western area, although its basal beds are similar to the basal Monterey of the central synclinal belt. Fossils in the latter area indicate the base of the unit to be Saucesian in the Paloma Creek-Atascadero vicinity but possibly as young as early Relizian near Trout Creek. The top of the unit also may transgress time, apparently ranging in age from late Miocene (early Mohnian or younger) near Atascadero to middle Miocene (Luisian) southeast of Trout Creek where the upper member is absent. The contact between the upper and lower member of the Monterey is Luisian to early Mohnian in age—probably late Luisian.

East of the Rinconada fault, the base of the Monterey ranges in age from early Saucesian north of Rocky Canyon to late Saucesian near the Salinas River. Only Relizian fossils were found in the Monterey southeast of there, and the unit presumably interfingers with the Vaqueros Formation in that direction. The upper contact of the unit is not dated, although the Santa Margarita Formation east of the fault contains megafossils (*Lyropecten estrellanus?, Ostrea titan*) suggestive of a late Miocene age.

Because of apparent differences in stratigraphic relations rather than age (see figure 6), the Monterey sequences on opposite sides of the Rinconada fault are believed not to have been deposited in their present relative positions. Therefore, large-scale strike-slip displacement along the fault seems likely.

The various Monterey sequences exposed in the map area are extensions of the Monterey Formation present in the surface and subsurface to the north and northwest. The lower calcareous member is correlative with the Sandholdt Member of the Monterey Shale to the north (Durham, 1968b; Smith and Durham, 1968), an unnamed lower shale member of the Monterey Shale to the east (Dibblee, 1968), and the Point Sal Formation to the southwest (Hall and

Table 6.   Miocene megafossils from localities in the Monterey and Santa Margarita Formations.[1]

Fossils	Localities[2]													
	SLO 5	SLO 60	SLO 61	SLO 84	SLO 85	SLO 88	SLO 90	SLO 92	SLO 125	SLO 126	SLO 138	SLO 139	SLO 175	SLO 183
ECHINOIDS														
Astrodapsis schencki Grant and Eaton										X				
whitneyi Remond										X				
sp.			X	X			X							
GASTROPODS														
Astraea biangulata (Gabb)									X					
Bulla cf. B. cantuaensis Anderson and Martin									X					
Crepidula sp.											X			
Forreria perelegans Nomland?											X			
sp.				X										
PELECYPODS														
Amussiopecten cf. A. vanvlecki (Arnold)	X													
Anadara cf. A. trilineata (Conrad)											X			
Anomia sp.					X									
Arca leptogrammica Hall											X			
Chione cf. C. temblorensis (Anderson)											X			
Chlamys cf. C. sespeensis Arnold														X
Crassostrea titan Conrad		X		X			?							
sp.									X					
Delectopecten cf. D. peckami (Gabb)												X		
?Dosinia sp.											X			
Florimetis biangulata (Carpenter)									X	X	X			
Hinnites giganteus (Gray)					X									
Leptopecten andersoni (Arnold)	X													
discus (Conrad)					X			cf					X	
Lucinisca n.sp.? cf. L. nuttalli (Conrad)											X			
Lyropecten estrellanus (Conrad)		X		?		X				X				
sp.									X					
Macoma sp.									?		X			
Miltha sanctaecrusis Arnold?														X
Mytilus schencki Hanna and Hertlein									X					
sp.											X			
Prototheca teherrima (Carpenter)											X			
Tellina sp.											X			
?Tresus sp.											X			
BARNACLES														
Balanus sp.				X						X				
Indicated age	Middle Miocene—"Temblor Stage"	Probably late Miocene	Late Miocene to Pliocene	Late Miocene	Late Miocene	Probably late Miocene	Late Miocene	Probably late Miocene	Late Miocene	Late Miocene	Late Miocene	Probably middle Miocene	Late Miocene	Early Miocene—"Vaqueros Stage"

[1] Fossils identified by W. O. Addicott, U.S. Geological Survey, Menlo Park.
[2] Monterey Formation—SLO 5, 138(?), 139, 183.
   Santa Margarita Formation—SLO 60, 61, 84, 85, 90, 92, 125, 126, 138(?), 175.

Table 7. Fossil diatoms from upper Miocene strata of the Monterey and Santa Margarita Formations. Identifications made by G Dallas Hanna, California Academy of Sciences, May 1967.

Flora	Sample localities		
	SLO 10[1]	SLO 52[1]	SLO 127[2]
Actinocyclus ehrenbergii	X		
sp.		X	
Actinoptychus grundlerii			X
senarius	X	X	X
splendens	X	X	
Arachnoidiscus ehrenbergii			X
ornatus		X	
Biddulphia (2 species)			X
Chaetoceras (several species)	X		
Coscinodiscus asteromphalus		X	X
eremanii		X	
excentricus			X
fulguralis	X		
lineatus	X	X	X
marginatus		X	X
perferatus			X
radiatus		X	
several species	X		
Denticula lauta	X	X	
Gephyria gigantea		X	
Hercotheca mammillaris	X		
Isthmia enervis	X		
Melosira clavigera			X
Navicula lyra		X	
longa	X		
Pterotheca sp.	X		
Rhaphoneis amphiceros	X		
Stephanopyxis appendiculata	X	X	
Triceratium montereyi		X	
Xanthiopyxis ovalis	X		

[1] Reported to be upper Miocene, but not quite as high as the top of the Monterey Formation near Monterey.
[2] Reported to be "a typical upper Monterey Miocene assemblage found widely distributed in California."

Corbato', 1967). The upper siliceous member essentially corresponds to the upper Monterey Shale of Durham, the Whiterock Bluff Member of the Monterey of Dibblee, and the Monterey Formation of Hall and Corbato'. The lower Monterey and associated mafic volcanic (and intrusive?) rocks have their time and lithologic equivalents to the northwest (Burch and Durham, 1970) and to the west and southwest (Page, 1970b; 1972).

Santa Margarita Formation

This is a poorly stratified, sandy, marine sequence that conformably overlies the Monterey Formation. The Santa Margarita Formation is confined to the synclinal trough between the Rinconada and "Nacimiento" fault zones, except for two small exposures east of the Rinconada fault in the northern part of the mapped area. The formation was named after the town of Santa Margarita (Fairbanks, 1904), although a type area or locality was not specified. Fairbanks included some of the Paso Robles Formation with the Santa Margarita For-

mation on his map. Richards (1933) studied the formation and its fossils and designated the discontinuous exposures along Santa Margarita Creek as the type section.

The Santa Margarita largely consists of thick beds of weakly consolidated, coarse, arkosic sandstone that is nearly white in outcrop. It grades laterally into fine to medium sandstone or conglomerate. Mudstone, siltstone, diatomite, and their silicified counterparts form interbeds, particularly in the lower part of the formation. The unit is commonly interspersed with resistant shell beds and reefs. The formation develops a gentle rolling terrain that is largely mantled with a thick sandy soil. Except in stream and artificial cuts, bedding attitudes are sparse.

Sandstone of the Santa Margarita Formation forms massive to coarsely cross-bedded strata as much as 20 feet thick. Most of the sandstone is pale gray and consists of fine to coarse, moderately to well-sorted, subangular to subround, loosely packed grains of quartz, plagioclase, K-feldspar, and composite grains of these minerals. Micas and heavy minerals are present in inconspicuous amounts. Except where cemented with calcite, the sandstone is very friable. Some of the sandstone is pebbly and lenses of massive pebble and small-cobble conglomerate are interspersed throughout the unit. Rounded clasts of colorful volcanic rocks, similar to those in the nearby older units, predominate, although granitic rocks, siliceous Monterey debris, colored chert, diabase, sandstone, and quartz from nearby formations are present locally.

Mudstone, siltstone, and impure diatomite, some of which are tuffaceous, form thinly bedded sequences generally less than 10 feet thick. In some places, these rocks are soft and unaltered, especially along Rinconada Creek. Mostly, however, they are altered to chert, porcelanite, and siliceous shale and mudstone that are similar to the siliceous rocks of the upper Monterey Formation.

Shell beds are interspersed at intervals throughout the Santa Margarita and locally are useful markers where outcrops are absent. These consist of shell-rich sandstone, sandy bioclastic limestone and reefs. The beds range from less than a foot to 20 or 30 feet thick and generally are well cemented. Ostrea titan, species of echinoids, and various pectinids, are the principal shell constituents, although a fairly diverse fauna is represented. Those forms sampled are identified in table 6. Other fossil localities are common, and these contain additional forms.

The contact between the sandy Santa Margarita and underlying siliceous Monterey is generally fairly sharp, although the two units are somewhat gradational by interbeds in most localities. One mile northwest of Trout Creek, an exceptionally thick (300-400 feet) transition zone exists between the two units. To the southeast of Trout Creek and also in the subsurface northeast of Santa Margarita, the Monterey is very thin and its upper member appears to be absent—probably because of lateral facies changes. A single known outcrop area of Santa

Margarita(?) sandstone, conglomerate, and limestone in the northeast part of the area rests depositionally on granitic rocks. The Paso Robles Formation generally rests with an angular unconformity on the Santa Margarita. However, in two places along Rinconada Creek, the Paso Robles(?) beds appear to be gradational and crudely conformable with the steeply dipping Santa Margarita beds. The thickness of the Santa Margarita varies from about 200 feet west of Atascadero to a maximum of roughly 2000 feet northeast of the town of Santa Margarita.

Abundant coarse clastics, shallow water marine fossils, large-scale cross-bedding, and lenticular beds indicate that most of the Santa Margarita Formation was deposited in a shallow marine, high-energy, nearshore environment. The finer clastic interbeds, however, suggest somewhat deeper or at least more sheltered conditions at times. The thick sequence of shallow marine sediments near Santa Margarita requires that deposition was concurrent with progressive local basinal subsidence. An eastern granitic source probably provided most of the sediment. However, the local presence of Franciscan and various Jurassic rocks in the unit, as well as abundant conglomerate and absence of upper Monterey beds in the southeastern part of the area, indicates a probable southeastern source. It is possible that the synclinal trough now defined by the Rinconada and "Nacimiento" faults had begun to form at least as early as late Miocene time. Relative uplift within or near the map area also is indicated by the presence of Monterey debris within the Santa Margarita unit. The presence of a few angular worm-bored clasts of porcelanite near the base of the Santa Margarita Formation a mile north of Santa Margarita suggests a nearby shoreline cut in the Monterey Formation. That the shoreline lay to the east is indicated by the arkosic sand, which must have come from the La Panza Range granitics.

A late Miocene age for the Santa Margarita is indicated by abundant fossils (tables 6 and 7). The base of the unit is no older than Mohnian (late Miocene) near Atascadero, based on foraminifers in the underlying Monterey Formation (SLO 169, table 5). However, the base of the Santa Margarita near Trout Creek could be as old as late Luisian (middle Miocene) (SLO 116, table 5). There is no direct evidence that any part of the Santa Margarita is any younger than late Miocene.

Paso Robles Formation

The Paso Robles Formation is a predominantly nonmarine conglomerate unit that is widely and nearly continuously distributed over 1000 square miles of the upper Salinas Valley area (Galehouse, 1967). It was named by Fairbanks (1898) for characteristic outcrops near the town of Paso Robles, about 10 miles north of Atascadero. In the mapped area, the Paso Robles is widely distributed along the Santa Margarita syncline and on the north flank of the La Panza Range. It is not present west of the "Nacimiento" fault zone. It overlies the Mon-

terey and Vaqueros Formations, the granitic rocks, and much of the Santa Margarita Formation with an angular unconformity. Locally, however, the Santa Margarita-Paso Robles contact appears to be conformable and may not represent a large gap in time. Older alluvium overlies the Paso Robles, and in places along the Rinconada fault zone the two units are extremely difficult to differentiate.

The formation consists of poorly exposed, weakly consolidated mixtures of gravel, sand, silt, and clay. The coarse clastics are dominant, forming lenticular beds of massive to cross-bedded conglomerate and sandstone. These show abundant fluvial features (scour and fill, large-scale crossbeds). Interbedded impure claystone, siltstone, and rare marl beds represent former pond and lake deposits. Much of the unit is poorly sorted. Debris from the Monterey Formation predominates and gives rise to a clayey soil and local caliche deposits. Other sources predominate locally. A distinctive basal conglomerate and sandstone unit can be recognized locally west of the Rinconada fault.

Poorly sorted conglomerate lenses, composed of Monterey Formation clasts set in a sandy, muddy matrix, characterize the formation. The clasts range in size from sand to boulders 5 or 6 feet across. These mainly consist of siliceous shale and mudstone, procelanite, chert, dolomite, sandy limestone and calcareous sandstone of the Monterey and Vaqueros Formations. Varying amounts of volcanic, granitic, sedimentary, mafic intrusive, and metamorphic detritus from other nearby units also are common locally. Depending on durability, initial clast shape, distance traveled, and amount of reworking, the clasts vary from angular to well rounded. Pholad-bored, large (to 5 feet long), angular boulders—mainly dolomite, calcareous sandstone, and limestone of the lower Monterey and Vaqueros units—are notable in the lower part of the Paso Robles, especially in the vicinity of Santa Margarita. These distinctive clasts lie within or just above the basal member of the Paso Robles. They could not have been transported very far and clearly are evidence of a former marine shoreline. Pholad-bored clasts also are present higher in the formation but are smaller, more rounded, and less prevalent.

Sandstone is rather abundant as interbeds both in the conglomerate and in the finer clastics. These are arkosic but contain varying amounts of Monterey and other debris. Sandstone composed entirely of granitic debris is dominant in the northeastern part of the area where the unit laps onto the granitic basement. Most of the sandstone is very friable and poorly to moderately sorted. Locally, however, it is well cemented with calcite.

Cream to light-yellowish or grayish-brown claystone and mudstone form well-bedded sequences with sandstone, siltstone, and minor conglomerate as much as 15 feet thick or more. These clayey units are poorly exposed and generally covered with a thin black clayey soil. The claystone contains variable amounts of sand, silt, and other impurities. In one sample, broken fragments (reworked) of marine diatoms were recognized by G. D. Hanna (1969, oral communication). Other fossils included scattered coal-like plant debris and rare thin-shelled mollusks of probable lake origin. A single x-ray diffraction analysis of a typical sample of impure diatomaceous claystone showed a mixture of montmorillonite, quartz, plagioclase, K-feldspar, and small amounts of cristobalite(?) and kaolinite. The diatoms presumably are composed of x-ray amorphous opal.

Calcareous beds and lenses, as much as 6 inches thick, are interbedded with the claystone. These rocks contain siliceous zones and concretions and presumably are a type of lake marl.

A basal member of the Paso Robles Formation is well developed at Chalk Hill and south of there, where it is thick enough to differentiate on the map (TQpb on plate 1). This unit consists of pebble and small-cobble conglomerate and interbedded coarse sandstone. It is identified by an abundance of relatively well-sorted, well-rounded clasts composed mainly of volcanic porphyry. A small minority of the volcanic and other gravel is angular to subangular, however. The sand beds and conglomerate matrix form a coarse, porous arkose almost free from silt and clay. Within the unit are large, angular, pholad-bored clasts. The unit overlies the Santa Margarita Formation, partly with an angular unconformity. It grades rapidly upward, partly by interbeds, into typical Paso Robles composed mainly of Monterey debris. The lower unit is about 100 feet thick due east of Santa Margarita but much thinner elsewhere. Except for re-worked(?) shells (generally worn or broken) of *Ostrea titan* at its base, no diagnostic fossils were noted in the unit. The basal member is believed to be mainly nonmarine in origin and composed largely of reworked debris from a former shoreline cut into the Monterey Formation during Santa Margarita time. However, marine fossils were reported from the base of the Paso Robles Formation just north of the mapped area (Addicott and Galehouse, 1973). The marine fossils are considered to be in-place and not reworked.

The Paso Robles formation has been largely removed by erosion from the Santa Margarita syncline, where it reaches a maximum thickness of about 300-400 feet. North of the La Panza Range, it attains a maximum thickness of about 500 feet. Common fluvial features indicate deposition as stream channel and flood plain deposits. Some of the large boulders, however, may have been deposited by debris flows. Clayey and marly lakebeds indicate local ponding, which may have resulted from tectonic warping or faulting. The sediment sources were local, and the main source areas and drainage patterns are believed to be similar to the present ones. Paleocurrent and heavy mineral data indicates that the Paso Robles streams flowed northward in this region (Galehouse, 1967).

Although diagnostic fossils were not identified in the study area within the Paso Robles unit, its unconformable position over part of the Santa Margarita Formation is suggestive of a post-Miocene age. Angular pholad-bored clasts are well preserved and indicate that post-Miocene weathering was not prolonged, however. Therefore,

the lowest Paso Robles beds probably are early Pliocene in age. The only diagnostic fossil reported from the Paso Robles is a fossil pinniped, apparently from the basal member a mile east-southeast of Santa Margarita (Kellogg, 1921). According to C.A. Repenning (1971, oral communication), this incomplete fossil (limbs) is from a seal or a walrus of Miocene or Pliocene age. Repenning reports the fossil bones to be unworn and in place. If so, the basal member presumably is at least partly marine and no younger than Pliocene. Extensive deformation of the Paso Robles, as well as the presence of older alluvium locally overlying it, suggests a minimum age of about early Pleistocene. The Paso Robles Formation of the upper Salinas Valley, including that in the area mapped, is considered to be early to late Pliocene in age by Galehouse (1967) and not Plio-Pleistocene or Pleistocene as suggested by most other investigators.

Older Alluvium

The older alluvium consists of weakly consolidated gravel, sand, silt, and clay in lenticular and interfingering beds laid down as flood plain, channel, fan, colluvium, and lake deposits. All stages of morphologic preservation can be recognized—from relatively young, slightly dissected terrace and fan deposits to presumably older deposits with no vestige of their original morphologic character. Some of the older alluvial deposits along the Rinconada fault exhibit disturbed bedding locally and appear to be faulted. Based on morphology, stratigraphy, and lithology, "relatively older" and "relatively younger" portions of the older alluvium were identified locally, but correlation of these subunits from one locality to another is not intended. Also, no effort was made to discriminate among the two or three terrace levels that can be recognized locally.

The composition of the older alluvium is variable, depending on whether it was derived from the La Panza Range (mainly granitics) or the Santa Lucia Range. A variety of rocks from both ranges are blended in some beds, indicating a more regional source. Some of the deposits are very thin, and most do not exceed 30 to 40 feet in thickness. The older alluvium(?) northwest of Fivemile Bridge appears to be about 200 feet thick and may be part of the Paso Robles Formation.

The older alluvium probably is Pleistocene in age. Based on comparative morphology, the slightly dissected, low-level terrace deposits are likely to be latest Pleistocene. Some of the higher and more poorly preserved units could be as old as early or middle Pleistocene. The only fossils identified (sample SLO 100) were obtained from relatively horizontal beds half a mile east of Onemile Bridge. The sample, containing freshwater ostrocods, small gastropods, and a few oogoniums of Chara, was identified as "Probably Pliocene or Pleistocene" by C.C. Church (1966, written communication).

The correspondence of compositions between the older alluvium and the adjacent modern stream deposits (younger alluvium) clearly indicates that the present drainage patterns have remained fairly constant since late Pleistocene time. However, "relatively older" deposits are not associated with present streams (or at least not with large streams). The thick alluvial and lake bed deposits along Pozo Road between Onemile and Fivemile Bridges suggest the existence of a former northwest drainage. The lake beds may have been deposited in depressions resulting from Pleistocene deformation.

Younger Alluvium

Modern stream channel, flood plain, and alluvial fan deposits are mapped as younger alluvium. Also included are very young, low-level terrace deposits that are at least locally subject to flooding. Most of these Holocene deposits are less than 10 feet thick, although they commonly overlie and grade downward into thicker older alluvium. Composition is variable, depending on whether the clasts and grains are locally or regionally derived. The Salinas River deposits are mostly sandy granitic debris, although the gravel is composed of a variety of rocks: granitic rocks, volcanic porphyry, and sandstones, with subordinate siliceous Monterey rocks, Franciscan debris, quartz, quartzite, aplite, serpentinite, and other rock types.

Minor pond deposits contain fine-grained organic sediment along Santa Margarita, Trout, and Rinconada Creeks and their tributaries. Such deposits and local small depressions may reflect minor subsidence within the Santa Margarita syncline or locally along faults.

Landslide Deposits

The principal landslide deposits lie in the Santa Lucia Range, being concentrated mainly within weak rocks of the Franciscan melange and Toro Formation and along fault zones. All landslide deposits are shown as undifferentiated Qls on plate 1, although a variety of landslide types and ages were observed. Many of the slides (or parts of them) have been historically active. Some landslide deposits, however, are clearly older, being relatively stable and partly modified or dissected by erosion (e.g., some of those east-southeast of Eagle Ranch and most of the large slide at McClappin Spring). All of the landslide terrain, however, should be considered potentially unstable. Active landslides were not distinguished from inactive landslides during this study because sliding tends to be intermittent and partly dependent on the amount of seasonal rainfall and on such longer term geologic phenomena as erosion and over-steepening of slopes. In some cases, landslides have been reactivated or newly created by man's activity. Fortunately, most development has occurred in the more stable areas.

Landslide deposits may be classified as rock falls, rotational landslides, block glides, and mud or debris flows. Most slides, however, are complex combinations of these and no attempt was made herein to classify them. The larger slides in par-

ticular are complex, and only parts of them may be active at a given time. Slides smaller than about 100 by 200 feet and areas of rapid soil creep (especially in the Franciscan) generally were not indicated on the map.

Many rapid mud or debris flows occurred during the winter of 1968-1969 when an abnormally large amount of rain fell. During this period, small debris flows (a few hundred tons or less) were initiated at the steep heads of gullies and small streams. Based on the sizes of clasts, steepness of slopes, slide-path markings, minor damage to structures and vegetation, and other factors, it is believed that some of these flows moved downhill as fast as 20-30 miles per hour. Upon reaching valleys and fans of gentle slope, the muddy matrix was washed away, leaving only a veneer of coarse debris as evidence of the event. Because the remaining debris was of no significant thickness, most of these deposits were not mapped. However, numerous scars were evident at the heads of many gullies in the Paso Robles, Monterey, and Atascadero Formations. The larger flows of this type can be very destructive, although only minor fence damage was observed in the mapped area. The largest recent debris flow noted carried clasts at least 8-10 feet across. This type of debris flow may occur infrequently at any given gully head, but it probably is a principal form of headward erosion in some parts of the area. It is believed that small alluvial fans at the foot of many small gullies are formed largely from the debris of these small debris flows. Dissected remnants of older debris slide deposits east of Eagle Ranch, containing slabs of Atascadero Formation sandstone as much as 20 feet long, testify to the high energy potential of rapid debris flow phenomena.

STRUCTURAL FEATURES

The principal structural features of the mapped area are the Rinconada fault, the "Nacimiento" fault zone, and the intervening Santa Margarita syncline. These and the associated less important features reflect at least two periods of prolonged tectonic disturbance; one or more during late Mesozoic-early Tertiary(?) time and one since mid-Tertiary time. The older features are difficult to document because of later obscuring tectonism. However, it seems probable that the principal structural features developed in association with and as a result of underthrusting of the North American continental plate by the former (now consumed) Farallon oceanic plate (McKenzie and Morgan, 1969). The later period of deformation is believed to be related to right-lateral strike-slip movement between major plates (Pacific and North American) during which time the greater San Andreas fault system developed, beginning around mid-Tertiary time. The rationale and nature of the tectonic framework of the California Coast Ranges is presented in more detail by McKenzie and Morgan (1969), Atwater (1970), Hamilton (1969), Page (1970a, 1970b) and others.

Some of the structural features and relations mapped by this writer (e.g., the Rinconada fault, relations between various upper Mesozoic units) generally support the recent concepts and models put forth by the above named investigators. However, the area is too small and complexly disturbed for most structural features to be directly attributed to major tectonic events.

FAULTS
Rinconada Fault

The Rinconada fault is a rather linear, narrow, near-vertical zone of faults that marks the western margin of the La Panza Range and locally defines the western margin of the Salinian block. It can be traced from its juncture with the "Nacimiento" fault zone just east of the mapped area for about 25 miles north-northwesterly to where it joins the San Marcos (Jolon) fault (Jennings, 1958). The southern 16-mile segment of the fault lies in the mapped area. The Rinconada fault probably is a segment of a greater zone of high-angle strike-slip faults extending 85 miles to the north which includes the Jolon, San Marcos, Espinosa, and Reliz faults.

Within the mapped area, the Rinconada fault zone is well defined over most of its length. It is difficult to identify only in alluviated areas. Geologically, it is recognized as one to several, closely spaced, parallel to branching faults that clearly truncate all pre-Quaternary units. The zone is generally marked by a single main break in any given area. The significance of the fault is indicated by the highly contrasting terranes which it separates. The granitic rocks and the Simmler Formation(?) are confined to the east side of the fault, whereas the Franciscan melange, Toro Formation, ultramafic-mafic complex, and mafic volcanic rocks lie only west of the fault. That the Rinconada fault (rather than the "Nacimiento" fault zone) is the local boundary between the granitic Salinian block and the Franciscan (Nacimiento) block is indicated by the Franciscan outcrops along the southern segment of the fault. This indication is supported by gravity anomalies to the northwest (Burch and Durham, 1970). Other contrasts in character, thickness, and/or age also can be seen in the Upper Cretaceous rocks and the Vaqueros and Monterey Formations which are exposed on both sides of the fault.

Physiographically, the Rinconada fault zone is expressed by an alignment of saddles, notches, drainages, and low scarps. Linear soil, vegetation, and water seepage zones also help to define the several fault strands, at various times of the year. Sheared and brecciated rock, zones of cementation and alteration, and locally steep and erratic dips help to identify the zone and its separate breaks locally. The relatively linear features strongly indicate that the fault zone and most of its separate fault breaks are vertical or nearly so. Near-vertical fault planes and shears observed in the zone support this indicated orientation.

Large-scale strike-slip movement along the fault zone is indicated by: 1) clockwise drag of bed-ding and stratigraphic units as the fault zone is approached; 2) slivers of exotic Franciscan rocks along the fault zone; 3) clockwise rotation of streams and ridges as the fault zone is approached; 4) apparent offsets in streams and ridges in several places; 5) horizontal grooves in soft sandstone of the Paso Robles Formation(?) in Rinconada Creek; and 6) sharp contrasts in geologic terranes on opposing sides of the fault.

Attempts to match offset units across the Rinconada fault within the map area were not successful. However, the rather unique Simmler(?)-Vaqueros sequences east of the fault—giant granitic boulder conglomerate and associated marine beds—can be correlated in age and lithology with the Tierra Redonda Formation of Durham (1968a) in the Harris Valley area 33 miles to the northwest. The correspondence between these separate outcrop areas, one east of and abutting the Rinconada fault and the other 1 or 2 miles west of the Jolon (San Marcos) fault, suggest right-lateral offset of about 33 miles since early Miocene time along the connecting Rinconada and Jolon faults. Other units northwest of the map area and west of the Jolon fault also appear to correlate generally with the lower member of the Vaqueros Formation and the unnamed Upper Cretaceous sandstone-conglomerate unit east of the Rinconada fault in the southeast part of the mapped area. Specific correlations, however, have not been documented. A suggested maximum right-lateral displacement of about 40 miles since Late Cretaceous or early Tertiary time, postulated by Dibblee (1972), are consistent with this writer's conclusions. An estimated right-lateral offset of about 11 miles since early Pliocene time along the fault northwest of Paso Robles (Durham, 1965) also supports this conclusion. If the rate of displacement along the Rinconada fault was more or less constant (about 1.75 miles/1,000,000 years) and the maximum offset was 40 miles, the fault may have had its inception during late Oligocene time (25-30 m.y. ago).

The fact that the Paso Robles Formation is clearly offset by the Rinconada fault, both within and beyond the mapped area, suggests that faulting continued through Pliocene and into Pleistocene time. Pleistocene displacement is also indicated by apparently offset and up-turned beds of older alluvium of probable late Pleistocene age. Late Pleistocene or Holocene activity is indicated by a sag(?) pond half a mile northwest of Fivemile Bridge, local offset and clockwise-rotated drainages(?), and a few less obvious depressions, wet spots, and faint aerial photo lineaments in younger alluvium in the Rinconada drainage area. Unequivocal offset of the younger alluvium has not been identified, however. Neither has any historic displacement been identified along fence lines or roads. Nevertheless, crudely located scattered seismic epicenters in the vicinity of the mapped area (California Department of Water Resources, 1964, plate 2) suggest possible activity along the Rinconada or nearby faults.

A number of relatively minor faults connect with or lie adjacent to the Rinconada and presumably resulted from the same right-lateral stress system that caused displacement along the Rinconada fault. Some of these faults splay out from the main Rinconada breaks and are likely to be right-lateral type faults. Others, however, appear to be normal and the result of extension; for example, the north- and northwest-striking faults 1 to 2 miles east and southeast of Fivemile Bridge.

"Nacimiento" Fault Zone

The northwest-trending "Nacimiento" fault zone is an ill-defined, complex array of faults of diverse types and ages. It lies on trend, both locally and regionally with faults and fault zones generally identified as the Nacimiento to the southeast (Hall and Corbato', 1966; Vedder and Brown, 1968) and the Nacimiento or Sur-Nacimiento to the northwest (Jennings, 1958; Page, 1970a). Although the southeast segment of the fault zone has been identified as the Rinconada fault (Dibblee, 1972), the latter appears to be truncated by southwest-dipping thrust fault of the "Nacimiento" fault zone at the eastern margin of the mapped area (herein; Eckel and others, 1941, plate 85). The above workers and others (Reed, 1933; Loney, 1970; Hill and others, 1958) give various and conflicting definitions, descriptions and interpretations of the Nacimiento fault zone and the nature and age of its movements. It seems clear that the Nacimiento fault zone is not a single master break of specific age, but rather a complexity of branching and discontinuous faults of diverse orientations, movements and ages. Therefore, the term "Nacimiento" is used in a modified way herein to identify this poorly defined fault zone of probable multiple origin.

In the mapped area, the "Nacimiento" fault zone is more or less defined by the narrow sinuous outcrop-band or "corridor" of Franciscan melange that separates rocks of the southern Santa Lucia block from those of the Santa Margarita syncline. Aside from the late Mesozoic to early Tertiary pervasive shears that are an integral part of the Franciscan melange, there are two main fault elements that comprise the "Nacimiento" fault zone: those that dip to the southwest and tend to be sinuous in outcrop trace and those that are presumably vertical, or nearly so, and have linear trends. Additionally, in places there are faults and fault segments that dip to the northeast and have other orientations. Some also diverge from the main trend.

Southwest-Dipping Faults

Southwest-dipping faults can be identified along the entire length of the fault zone, particularly along the southwest margin of the Franciscan melange "corridor" but also along its northeast margin. In places, two or more subparallel faults dip to the southwest, displaying an over-all imbricate pattern. Dips vary from horizontal to near vertical but generally are 30-75° SW. The fault planes and their specific characteristics are rarely exposed, and most attitudes must be interpreted from the sinuous fault traces in conjunction with topographic expression. Overturned and truncated beds of the Monterey and Santa Margarita Formations on the northeast side of the fault zone provide local clues as to the nature and time of faulting. The best exposure of southwest-dipping faults and shears is in the 200-foot-long spillway—a cut trench with N30°E trend—on the north side of the reservoir located 1-1/4 miles northwest of Santa Margarita and about 1,000 feet east of the highway. Here, the overturned and broken Monterey-Santa Margarita sequence is in fault contact with sheared serpentinite and associated fault slices of Franciscan melange and Atascadero Formation. The main fault contact dips 75° SW; the subsidiary shears on either side dip from 25° SW to 80° NE. A similar relationship between Franciscan and disrupted and overturned Miocene strata is not so well exposed in Santa Margarita Creek 4,000 feet to the southeast. Clearly the structural features indicate a strong component of reverse dip-slip movement since late Miocene time with the southwest block up in this area. Similar structural-stratigraphic relations can be readily interpreted elsewhere along the fault zone (e.g., northwest of Devils Gap, northeast and east of Eagle Ranch, and southeast of Burrito Creek), suggesting that reverse and thrust displacement was widespread along the "Nacimiento" fault zone. Other southwest-dipping faults, mainly along the southwest margin of the fault zone, are subparallel to the above faults and may have similar reverse movement, although they do not truncate Tertiary rocks.

Some of the southwest-dipping faults along the southwest margin of the fault zone in the Atascadero quadrangle connect with or truncate folded thrust faults that wander off of the main fault trend. Presumably the folded faults are associated with an earlier period of thrusting (see Page, 1972, figure 3), the thrust sheets being deformed by the later disturbance along the "Nacimiento" fault zone. The cross-sections (plate 2) show the inferred relations across the fault zone. The magnitude of displacement along the southwest-dipping faults generally cannot be determined. However, overturning and truncation of Miocene beds in the vicinity of Santa Margarita (C-C' and D-D' in cross-sections) indicate a minimum dip-slip displacement of about half a mile. This displacement is associated with compression and apparent crustal shortening in a northeast-southwest direction.

The important southwest-dipping faults, although not unique to the map area, are in sharp contrast with northeast-dipping faults commonly reported elsewhere along the so-called Sur-Nacimiento fault zone (Vedder and Brown, 1968; Page, 1970a; Gilbert, 1971).

Vertical Faults

Numerous northwest-trending vertical and near-vertical faults, 3 miles long or less and commonly in subparallel, branching or en echelon groups, are present in several places along the

"Nacimiento" fault zone and locally branch from it at acute angles. These faults, identified by their linear to gently curved traces, are most common within and along the northeast side of the fault zone. Many of the linear faults cut upper and middle Miocene rocks and some truncate southwest-dipping faults (some of which are late Miocene or younger). Locally, however, the linear faults appear to be truncated by or to merge with the late Miocene or younger non-vertical faults. The data indicate that many of the vertical faults formed during late Miocene time or later—at least partly contemporaneously with reverse faulting (i.e., northeast-southwest crustal shortening). A small depression (sag pond?) near W 1/4 cor. sec. 18 (proj.), T. 30 S., R. 14 E., suggests late Quaternary deformation, although the Quaternary alluvium is not clearly offset.

Direct evidence for the sense of movement along the numerous linear faults is absent and locally one can make a case for either dip-slip or strike-slip (both right and left) displacement. Perhaps both types of displacements have occurred. However, the presence of exotic fault slices, near vertical dips and tight folds in the lower Atascadero and Miocene units suggest large scale displacement—particularly in the Lopez Mountain quadrangle—that is most apt to be accommodated by strike-slip movement. Relations across many faults cannot be described simply in terms of dip-slip movement unless two or more periods of reversed movement are invoked. The relatively short lengths of the linear fault traces mapped does not necessarily argue against strike slip movement, as such faulting may have been transferred to branching or nearby *en echelon* faults.

A related zone of high-angle faults does appear to extend from near Burrito Creek to Atascadero Creek, a distance of about 12 miles; and its traces may continue northwestward within the Atascadero or lower Monterey Formations where bedding is disturbed. The presence of the contemporaneous(?), subparallel Rinconada fault—a major right-lateral fault—suggests that the linear faults of the "Nacimiento" zone may be directly related to the Rinconada stress system. In fact, the linear "Nacimiento" faults appear to branch from the Rinconada fault in the southeast part of the map area. Right-lateral strike-slip displacement appears to be more compatible with contemporaneous reverse faulting along the "Nacimiento" fault zone than does normal or left-lateral faulting. Apparent alternate compression and extension in a northeast direction may be the result of a common regional stress system in which right-lateral displacement is distributed over a broad zone of multiple northwest-trending faults, including the Rinconada, parts of the "Nacimiento", and unspecified faults to the southwest. Large scale displacement would create secondary stresses and structures (including reverse, thrust and normal faults of varying orientations, as well as folds) as the system of heterogeneous fault blocks was progressively adjusted and reoriented.

Other Fault Elements of the "Nacimiento" Zone

In addition to the principal southeast-dipping and vertical faults, there are other types of faults within the "Nacimiento" zone. For example, the mile-long segment on the northeast flank of the zone, where it crosses Atascadero Creek, dips moderately to steeply northeast with Atascadero Formation overlying Franciscan. This segment may be a remnant of the sole of a thrust plate along which the Atascadero Formation overrode the Franciscan, possibly during late Cretaceous or early Tertiary time (Page, 1970a; 1970b). Small, folded, thrust-plate remnants and other fault blocks of upper Mesozoic rocks also rest on the Franciscan west of Paradise Valley, northeast of Eagle Ranch, between Burrito and Trout Creeks, and elsewhere within the "Nacimiento" zone. Some of these may be pre-Miocene fault blocks, whereas others may be imbricated slices of the younger southwest-dipping fault elements.

The salient of Franciscan melange that projects northward from the main zone northeast of Eagle Ranch is peculiar and not well understood. The east side of the salient appears to be a northern extension of a southwest-dipping thrust or reverse fault. The fault on the west flank of the salient appears to dip steeply and may be more complicated than shown on the map. This and the west-trending intersecting fault to the north can be interpreted as having both dip-slip and strike-slip components with the east side apparently moving up and northwestward. Perhaps the Franciscan salient is a small diapiric structure. Two smaller isolated Franciscan outcrop areas, northwest of Santa Margarita and southwest of Trout Creek, also have uncertain structural relations, except that both are partly bounded by southwest-dipping faults.

Other Faults

Faults West of the "Nacimiento" Fault Zone

The several fault strands that lie southwest of the main "Nacimiento" fault zone probably are in part older thrust faults formed during earlier periods of westward or southwestward overthrusting and/or gravity sliding. These faults are partly folded and truncated by deformation along the "Nacimiento" zone (plate 1, cross-section C-C'). A fault along upper Atascadero Creek southwest of Eagle Peak is rather linear and may be nearly vertical. It extends to the southeast beyond the map area where it and related branching faults cut rocks of Miocene age (Page, 1972, figure 3).

Faults Between the "Nacimiento" and Rinconada Faults

Several relatively minor faults lie between the two major fault zones. The oldest lie entirely within the Atascadero Formation and may be partly the result of Late Cretaceous gravity sliding as suggested by internal deformation within this unit. Other faults are mainly linear and trend northwestward, a few appearing to interconnect with the Rinconada and its branches. The youngest of these truncate the Paso Robles Formation east and southeast of Atascadero and are late Pliocene or younger.

Faults East of the Rinconada Fault

A large number of faults were identified in the field and on aerial photographs with varying degrees of confidence. Many of these appear to be relatively minor faults that splay out from the Rinconada fault and die out to the east. Associated with the branching faults are several north-trending normal faults. The most important of these separates unnamed Upper Cretaceous beds from the Vaqueros Formation in the vicinity of Las Pilitas Road. Vertical separation appears to be at least several hundred feet (west side down). The southern end of the fault appears to be offset right laterally by branches of the Rinconada fault.

Middle Branch Fault

Several linear faults were identified along the Middle Branch of Huerhuero Creek and constitute a northwest-trending fracture zone here called the Middle Branch fault. Although there is some uncertainty how the several fault strands connect, topographic lineations and sheared and shattered granitic rocks serve to identify the zone and to suggest its general vertical orientation. Truncation of the Santa Margarita(?) and Paso Robles Formations by some of the fault traces indicates the fault to be relatively young. The presence of several small lakes and depressions, such as Silver and Clear Lakes, along the fault may also indicate youthful movement, although the depressions do not appear to be sag ponds. They may be fault related, however, in that the depressions are situated in the mouths of minor tributaries. It appears that the depressions are the result of differential rates of aggradation, the middle branch of the Huerhuero carrying abundant sediment during peak floods whereas the minor tributaries carry little or no sediment. Thus, the main alluvial bed is built upward with deltaic upstream deposition in the minor tributaries. The minor tributaries are abundant and may have been created by stream offset and beheading due to strike-slip faulting. The clockwise rotation of most small tributaries and ridges in the vicinity of the Middle Branch fault and the parallel relations to the Rinconada and other major linear faults of the region suggest that right-lateral strike-slip movement was likely. The apparent minor offsets of the Paso Robles Formation and lack of evidence for tracing any of the branches northward indicate relatively minor displacement, at least since Paso Robles time. The Middle Branch fault is

shown to split into two branches to the southeast beyond the map area, the easterly branch connecting with the Huerhuero (La Panza) fault (Dibblee, 1968).

A poorly defined northwest-trending fault zone southwest of the Middle Branch fault probably constitutes a minor fracture system connecting the Middle Branch fault just east of the mapped area with the Rinconada fault at the north margin of the mapped area. Other minor faults and fractures of different orientations and presumably varying types of displacement can be identified throughout the granitic terrane. All of the faults east of the Rinconada fault appear to be relatively minor and may be secondary shear and fracture zones created within the granitic block by the more important right-lateral Rinconada and Huerhuero faults that bound the block.

FOLDS
Santa Margarita Syncline

This complex synclinal structure, or synform, lies between the "Nacimiento" and Rinconada fault zones, extending northwestward from the junction of these two faults at least as far as Atascadero. As shown on the map and cross-sections B-B' to F-F' (plate 1), it is complicated by other folds and faults. In addition, its east margin is locally truncated by the Rinconada fault and the western margin is locally overturned. This troughlike feature probably had its inception about late Miocene time when marine conditions changed from deep water (Monterey Formation) to shallow shelf conditions (Santa Margarita Formation). The eastern margin of the syncline probably was defined by the emerging granitic rocks of the La Panza Range. Uplift along the western margin and principal definition of the synform must have been post-Miocene as the Santa Margarita Formation is strongly folded. Confinement of the Paso Robles to the syncline area suggests that it became a depositional trough during Paso Robles time (Pliocene?). Locally folded and faulted Paso Robles beds reveal that later deformation, possibly Quaternary, was superimposed on the synform.

Other Folds

Outside the Santa Margarita syncline complex, the only significant fold observed is the syncline near Willow Spring in the western part of the area. Folding not only is reflected in the Miocene rocks but also is partly defined by the folded thrust-sheet of Atascadero Formation rocks. The time of folding cannot be closely dated but may be coincidental with the main late Tertiary deformation and northeast-southwest crustal shortening along the "Nacimiento" fault zone. There are many additional folds, but these are too small or too complex to indicate on the map.

SUMMARY OF GEOLOGIC HISTORY

The most important historic events identified or reflected in the study area are summarized below. It is believed that the rocks and structures were formed under a complex sequence of overlapping and interrelated events that produced constantly changing, prolonged to episodic conditions. These events presumably resulted from the tectonic interaction of oceanic and continental plates. In order to reconstruct historic events, one must look beyond the study area and also appeal to recently formulated concepts of plate tectonics, which are still subject to modification.

LATE MESOZOIC (JURASSIC-CRETACEOUS) TIME

1. Formation of upper mantle and crust (serpentinite, ultramafic-mafic complex, mafic volcanic rocks) at mid-ocean rise. Subsequent spreading of oceanic plates away from rise with concurrent deposition of radiolarian ooze (chert) under deep water conditions.

2. Concurrent deposition of Franciscan sandstone and shale in a deep trench bordering an eastern landmass; contemporaneous underthrusting of the landmass (North American plate) by the oceanic plate from west to east; subduction of trench and oceanic deposits beneath the continental plate causing increased crushing, shearing, mixing, and metamorphism of rocks with depth.

3. Concurrent development of magmatic arc to the east with eruption of volcanic rocks; intrusion of igneous rocks (granodiorite and adamellite) into older marine sedimentary rocks (metamorphism) and its volcanic cover. Uplift and erosion with progressive removal of volcanic-sedimentary cover and exposure of granitic-metamorphic "core", which supplied sediment of changing content for the Franciscan and Great Valley-type rocks.

4. Concurrent deposition of the marine Great Valley-type rocks (Toro and Atascadero Formations, unnamed Upper Cretaceous rocks) between the trench (to the west) and magmatic arc (to the east) on the continental shelf and slope.

LATEST CRETACEOUS TO MID-TERTIARY TIME

Possible time of major over-thrusting (east to west?), particularly with oceanic upper mantle-crust rocks and Great Valley-type rock thrust over Franciscan.

OLIGOCENE(?) TIME

Local consumption of the Farallon plate as the Pacific mid-ocean rise and trench systems initially encounter each other; collision of offset rise segments and trench cause a major reorientation of stresses and relative movement of the crustal plates (Atwater, 1970).

A. Creation of or renewed movement along the San Andreas fault as a transform fault; large-scale right-lateral displacement of a portion of the former magmatic arc (previously a southern extension of the Sierra Nevada) creating the Salinian fault block which now, along with Franciscan basement rocks to the west, is attached to the Pacific oceanic plate.

B. Initial development of the Rinconada fault, possibly as part of the San Andreas system; right-lateral offset of the ancient Coast Range thrust (Bailey and others, 1970) that separated Franciscan and Salinian basement rocks.

C. Interaction between offset rise-segments and trench creates complex tectonic relations along continental margin, forming small mobile basins and ridges; local emergence of "highs" in La Panza and Santa Lucia Ranges.

LATE CENOZOIC TIME

1. Deposition of nonmarine sediments (Simmler Formation and other units) on granitic rocks and Upper Cretaceous sedimentary units during Oligocene(?)-early Miocene time.

2. Progressive subsidence with deposition of conformable sequence of Miocene marine sedimentary beds (Vaqueros, Monterey, Santa Margarita Formations); some volcanism (Monterey Formation).

3. Filling and uplift of marine basins accompanied by faulting and folding at the end of Miocene time. Some east-west crustal shortening as evidenced by southwest-dipping reverse faults of "Nacimiento" fault zone and formation of Santa Margarita syncline during late Miocene and/or Pliocene.

4. Pliocene and early Pleistocene(?) nonmarine sedimentation (Paso Robles Formation), in local basins following or concurrent with deformation and uplift.

5. Nonmarine deposition (alluvium) in small drainage basins during middle and late Quaternary time.

6. Continued right-lateral offset along Rinconada fault and other deformation throughout Quaternary time, probably diminishing with time; probable minor faulting and other deformation (scattered earthquake epicenters in region) during historic time.

MINERAL RESOURCES

The developed mineral deposits and prospects are shown on the geologic map and briefly described in table 8. Exploratory wells drilled for oil and gas are listed in table 9. The only mineral resources known to be of economic interest (listed in approximate order of decreasing importance) are crushed and broken stone, sand and gravel, limestone(?) and manganese. Only crushed and

Table 8.   Mineral deposits and prospects. (See plate 1 for precise locations.)

Map no.	Name	Location (Sec-T-R)	Owner	Geology	Remarks, references, date of field examination
1		20-28S-12E (proj.)		Hard fractured dark mudstone with thin muddy sandstone interbeds; lower Atascadero Formation (unit 1).	Small irregular quarry; probable source of road surfacing material and fill; intermittently active (1968).
2	Bluebird mine	30-28S-12E (proj.)	J. R. Davis, Atascadero	Black manganiferous chert associated with greenstone; Franciscan melange. Ore on small dump is low-grade siliceous manganese oxide with veinlets of neotocite(?).	Developed by open cut and 2 short adits; about 100 tons of ore produced pre-1918(?) (Trask, 1950, p. 229). Inactive (1970).
3		29-28S-12E (proj.)	J. R. Davis, Atascadero	Hard, dark gray to olive shale-mudstone and sandstone; lower Atascadero Formation.	Moderate-sized hillside quarry; source of road surfacing and fill(?) materials; active recent years (1970).
4		27-28S-12E (proj.)		Sand and gravel from channel of creek; young alluvium.	Material scraped from creek bottom; probable source of road base and fill; active recent years (1968).
5		26-28S-12E (proj.)	State of California(?)	Massive to cross-bedded, weakly consolidated coarse sand and fine gravel; Santa Margarita Formation.	Large hillside quarry and pit; source of road subbase and fill for nearby U.S. Highway 101; inactive (1968).
6		36-28S-12E (proj.)		Cherty and siliceous shale of upper Monterey Formation.	Small quarry; source of road base or fill; inactive (1968).
7		36-28S-12E (proj.)	Eagle Ranch	Chert, porcelanite, shale, and associated rocks of the upper Monterey Formation; thin bedded, folded, and fractured.	Large shallow quarry; source of road base or fill; inactive (1969).
8		2-29S-12E (proj.)	Eagle Ranch	Sheared and fractured greenstone and chert; Franciscan melange.	Tiny quarry; minor source of crushed stone; inactive (1968).
9		19-28S-13E	Heilmann Ranch, Atascadero	Decomposed granitic rock.	Small quarry; source of road material(?); inactive (1967).
10		24-28S-13E	Chandler(?)	Basal sandstone and conglomerate of the Santa Margarita Formation(?) on granitic rock.	Prospect shaft (gold?) 12–15 feet deep; inactive (1969).
11		25-28S-13E		Reddish-brown gravel and sand; older alluvium.	Moderate-sized, irregular borrow-pit; inactive (1968).
12		31-28S-13E		Reddish-brown gravel and sand; older alluvium (terrace deposit).	Small, shallow borrow-pit; inactive (1968).
13		32-28S-13E	Elbert Gifford, Santa Margarita	Basal conglomerate of Vaqueros Formation on granitic rock.	10- to 12-foot deep prospect shaft (gold?); inactive (1966).
14		7-29S-13E	Prof. Bolay(?)	White, very friable arkosic sand with local iron stains; Santa Margarita Formation; possible source of glass and specialty sand.	Several small test pits (now filled) dug in 1960s to obtain data on sand composition; inactive (1968).
15		9-29S-13E	Ellis Anderson, Santa Margarita	Reddish iron-stained fractures in decomposed granitic rock.	Prospect trench 10–15 feet deep by 40–50 feet long; probably prospected for mercury (none seen); inactive (1966).
16	Kaiser-Santa Margarita (also Roselip; Southern Pacific quarry)	10-29S-13E	Kaiser Sand and Gravel Corp., Oakland	Porphyritic granodiorite and quartz monzonite; more or less weathered and decomposed in upper 50+ feet with hard fresh rock below; cut by numerous fractures and small faults (shear zones).	Large, irregularly benched quarry with total relief of about 300 feet; intermittent source of crushed stone (ballast, concrete aggregate, road base) and riprap since early 1920s. Acquired by Kaiser about 1969; installed new crushing-screening plant and greatly enlarged quarry and production to become principal source of high-quality aggregate in county; Laizure (1925, p. 537); Franke (1935, p. 458); active (1972).

...continued on page following

Table 8. *Mineral deposits and prospects. (See plate 1 for precise locations.)* -(continued)

Map no.	Name	Location (Sec-T-R)	Owner	Geology	Remarks, references, date of field examination
17		9-29S-13E	Perry Ranch (mainly)	Sand and fine gravel (younger alluvium) from channel and flood plain of Salinas River. Gravel mostly granitic and volcanic rocks with sandstone, quartzite, quartz, and chert; sand mostly quartz and feldspars. Deposit shallow and partly depleted.	Several shallow pits; former source of fine aggregate and other uses; inactive (1970).
18		11-29S-13E		Partly decomposed granitic rock.	Small quarry; probably used for local road construction; inactive (1967).
19		15-29S-13E		Decomposed and weathered granitic rock.	Three quarries ranging from 200 feet in diameter (smallest) to about 600 feet long (largest); probably used for county roads in 1950s and 1960s; inactive (1966).
20		15-29S-13E		Sand and gravel from channel and flood plain; younger alluvium; deposit shallow. Gravel consists of volcanics, granitics, chert (2 kinds), sandstone, quartzite, diabase, greenstone, serpentinite.	Small pits; partly used as source of concrete aggregate in early 1960s and prior by A. O. Johnson (J. Sturgeon, 1972, oral communication); inactive (1966).
21		17-29S-13E (proj.)	Santa Margarita Land and Cattle Co., San Francisco	Nearly white, massive, coarse, friable, arkosic sandstone with some fossil shell debris; Santa Margarita Formation.	Prospect cut 200 feet long with maximum 15-foot face; specialty sand prospect tested by Owens-Illinois in 1955; inactive (1966).
22		25-29S-13E		Pegmatite dike in quartz monzonite. Dike is less than 1 foot thick, nearly vertical and strikes north. It is composed of large crystals (partly intergrown) of K-feldspar, plagioclase and quartz with some weathered biotite "books" to 2 inches across.	Prospect pit about 10 feet in diameter and 3 feet deep; no known production; inactive (1969).
23	Santa Margarita Ranch shell deposit	28-29S-13E (proj.)	Santa Margarita Land and Cattle Co., San Francisco	Sandy coquina beds and oyster shell reefs 10 feet thick or more in arkosic sandstone; Santa Margarita Formation.	Developed by 200-foot-long bench cut with 12-foot face and several small prospect cuts nearby. No record of production, but at least 1000 tons of shelly material quarried—possibly for lime or limestone use. Leased by Comar Shell Co. until 1928 (Hart, in press); inactive (1966).
24		35-29S-13E	Santa Margarita Land and Cattle Co., San Francisco	Gravel deposit in older alluvium that dips 10–20° SW. Pebbles and small cobbles of volcanic, granitic, and other rocks with abundant sandy matrix; weakly consolidated.	Pit 200 x 300 x 12 feet; probable source of road base and borrow for local roads; inactive (1966).

broken stone and sand and gravel are produced at the present time. Specialty sands may be of significant economic value in the future. The potential for developing other mineral resources are minimal, although impure deposits are known (e.g., impure dolomite, limestone) and geologic host or source rocks exist (e.g., serpentinite for chromite, asbestos, mercury; chert for manganese, Monterey Formation for oil and gas) which could provide unforeseen resources at some future date. Known and potential mineral resources are discussed briefly below.

CARBONATE ROCKS

Limestone and dolomite beds are common in the Vaqueros, Monterey, and Santa Margarita Formations. The "crystalline limestone of brownish color... 2 to 3 miles north of Santa Margarita" (Logan, 1919, p. 689) may refer to impure algal limestone lenses in the lower Monterey Formation about 2 miles east-northeast of Santa Margarita. Massive, sandy shell beds of the Santa Margarita Formation 2 miles southeast of Santa Margarita constitute the only limestone deposit known to be developed (see map no. 23, table 8). The calcareous and dolomitic beds of the lower Monterey appear to be too impure to be of significant economic value, although they are present in large amounts.

CLAY

Clay deposits of varied composition are present in the Paso Robles Formation near its base along Santa Margarita Creek and in Quaternary alluvial units. Most of these are impure (silty, sandy) and occur as lenticular beds less than 10 to 20 feet thick. It is doubtful if any of these alluvial clay deposits would have any special use. Brittle shale and mudstone of the upper Monterey Formation, on the other hand, are rather absorbant and may have some use as cat litter, floor sweepings, and similar uses. Large deposits of such material were not noted but may exist.

MANGANESE

About 100 tons of manganese ore was produced at the Bluebird mine prior to 1918 (map no. 2, table 8). No other prospect is known although chert beds

of the Franciscan melange and mafic volcanic unit locally are stained with black manganese oxide along fractures.

MINERAL SPRINGS

Two groups of mineral springs, referred to as the Huer-Huero Springs, are reported 1.5 and 3.5 miles south of the town of Creston and should lie in the northeast corner of the mapped area. According to Logan (1919, p. 693), the northern group is "sulphuretted, cold and carries some iron." The southern group flows "2 to 3 gallons per minute of 'iron' water strongly sulphuretted." Neither group was identified in the field. Springs smelling of hydrogen sulfide, however, were observed along upper Paloma Creek about a mile east of the Eagle Ranch headquarters. Many other springs are present in the mapped area, but none was observed to be notably mineral bearing or warm.

PETROLEUM

At least eight exploratory wells were drilled for oil in the mapped area (table 9). None of these was

Table 9. Exploratory wells drilled unsuccessfully for petroleum, northern San Luis Obispo 15-minute quadrangle, California.

Map designation[1]	Company and well	Location (Sec.-T.-R.)	Year drilled	Total depth (feet)	Geologic data[2] (depth in feet)
A	S. E. Hogue (Rocket Petr. Co.); "Stratton" 1	26-28S-12E	1949	1175	Santa Margarita Formation 0–190 Monterey Formation 190–1117(?) Upper 190–850(?) Lower 850(?)–1117(?) Basal sand ("Vaqueros") 1072–1117(?) Upper Cretaceous(?) 1117(?) to bottom
B	Mar-Roo Syndicate; "O'Reilly" 1	26-28S-12E	1949	1835	Santa Margarita Formation 0–100 Monterey Formation 100–1482 Upper 100–568 Lower 568–1482 Basal sand ("Vaqueros") 1450–1482 Upper Cretaceous 1482 to bottom
C	Mar-Roo Syndicate; "Lampman" 1	36-28S-12E	1949–50	3011	Monterey Formation 0–1290(?) Upper 0–450(?) Lower 450(?)–1290(?) Upper Cretaceous 1290(?) to bottom
D	C. J. Schilling Co.; Well no. 1	1-29S-12E	1934	1800	Santa Margarita Formation at surface; reported bottomed in Cretaceous(?).
E	Santa Lucia Oil Co.; "Santa Lucia" 1	1-29S-12E	1950	697	Monterey Formation 0–697.
F	Petroleum Holding Corp.; "Pago Grande" 1	2-29S-12E	1949	1954	Monterey Formation 0–1563 Lower (? beds overturned) 0–90 Upper 90–526 Lower (repeated) 526–1563 Basal sand ("Vaqueros") 1517–1563 Upper Cretaceous(?) 1563 to bottom
G	W. D. Reis; "Santa Margarita" 1	16-29S-13E	1965	1951	Santa Margarita Formation 0–1951; may convert to water well.
H	Santa Margarita Land and Cattle Co.; "China Garden" 1	16-29S-13E	1948	2255	Older alluvium 0–10 Santa Margarita Formation 10–1982 Transition zone 1470–1982 Monterey Formation 1982–2105 Vaqueros Formation 2105–2154 Upper Cretaceous(?) 2154 to bottom

[1] Most wells only approximately located on map.
[2] Geologic data largely interpreted by author from well records.

successful, although minor oil shows or tar in fractures were reported in the Monterey Formation in at least three wells. An unspecified well south of Atascadero near El Camino Real, probably Mar-Roo Syndicate's "O'Reilly" 1 or "Lampman" 1, is said to have flowed a small amount of oil, but this could not be verified. A very old well, possibly a water well, on the Eagle Ranch is reported (Meridith Gates, 1968, personal communication) to have supplied a small amount of gas to light the old headquarters house during the 1800s.

In spite of the abundant organic source rocks of the Monterey Formation, large reservoirs of oil or gas are not apt to exist in the mapped area because: 1) most of the Monterey is overlain by relatively porous beds exposed at the surface; 2) the Tertiary strata have been largely disrupted by faults; and 3) only minor anticlinal structures exist. Presumably, most of the petroleum originating in the Monterey beds has already been "leaked" to the surface. The most likely reservoir beds appear to be sands of the lower Monterey, which are best developed in the Atascadero area.

SAND AND GRAVEL

Sand and gravel has been excavated in the mapped area from recent channel and flood plain deposits (younger alluvium) and from parts of the Santa Margarita Formation. Such deposits were used for a variety of construction purposes depending mainly on clast size, durability, hardness and composition. The principal remaining sources of concrete aggregate are the alluvial deposits of the Salinas River, although most of the gravel would have to be beneficiated to remove excess percentages of soft or reactive rocks. Channel and flood plain deposits of that river, although large, are excessively sandy in most places and contain little gravel. Terrace deposits contain more gravel but are smaller. Composition of the terrace gravels, formerly excavated for concrete aggregate 3 miles north of Atascadero by Walter B. Roselip, was given by Goldman (1968, p. 33).

Sand and gravel for use as bituminous aggregate and road-base materials may be of slightly lower quality than that used as concrete aggregate; and adequate sources of the former exist in the alluvium of the Salinas River, Huerhuero Creek, and Atascadero Creek. Subbase and fill material may be of lesser quality and can be obtained from many alluvial deposits, as well as from the Santa Margarita Formation. Alluvium from the Santa Margarita, Trout, and Rinconada Creek drainages tends to be excessively fine but probably includes small local deposits of older channel gravels. The basal part of the Paso Robles Formation also may be a useful source of gravel for road base and other construction materials.

SPECIALTY SAND

Nearly unconsolidated, white to pale buff, feldspathic sand of the Santa Margarita and lower Monterey Formations, present in several places in the map area, may have some value for the manufacture of glass and ceramic materials. Two samples from a bench cut in the Santa Margarita Formation half a mile northeast of Santa Margarita in sec. 17 (proj.), T. 29 S., R. 13 E. (map no. 21, table 8) were tested in 1955 by a major glass manufacturer. The coarse, massive, friable sand, composed mainly of quartz and feldspars, was beneficiated in a laboratory and evaluated as potential flint-glass raw material. After scrubbing and washing, the sand samples showed a partial analysis of 0.17% and 0.15% Fe_2O_3 and 5.4% and 5.6% $CaCO_2$ (shell fragments). Scrub losses (fines) of 14.0% and 18.9% and screen analyses suggest the feldspars to be partly decomposed. After magnetic separation (8.1% and 7.2% rejects), the Fe_2O_3 content was reduced to 0.13% in both sand samples. Heavy media separation revealed practically no heavy minerals, and the iron content was reduced no further. Although these analyses indicate an iron content too high for flint-glass use (a maximum of about 0.04% Fe_2O_3 is desired), the sand may be useful for colored glass, fiberglass, and various ceramic purposes (see Gay, 1957; Hart, 1966, p. 90-98, 106). Somewhat similar sand, but without notable fossil debris, was also tested more recently (results unknown) 3 miles north-northwest of Santa Margarita (map no. 14). It is believed that the Santa Margarita sands can be processed to obtain one or more economic end-products by a combination of beneficiation techniques, including washing, scrubbing, magnetic separation, froth flotation, and selective screening.

The above analyses are not necessarily typical of the Santa Margarita Formation, as the sands vary somewhat in grain size, composition, and degree of induration from place to place. As a result, the sands should vary somewhat in their amenability to beneficiation. For example, soft, fine- to medium-grained, well-sorted, massive, pale gray sand that would be amenable to processing was noted in a creek in NW 1/4 sec. 28 (proj.), T. 29 S., R. 13 E. This sand is lightly cemented with clay and readily disaggregates in water. The thick sand unit in the upper part of the lower Monterey Formation west of Atascadero also may be of economic quality, but it is poorly exposed and lies in an area that is largely subdivided and developed residentially.

STONE

Crushed and broken stone has been obtained from several formations for a variety of local and regional uses, especially for road construction. High-quality aggregate, riprap, and road base have come mostly from granitic rocks of the Santa Margarita quarry of Kaiser Sand and Gravel Corp. Since 1969, this quarry has been greatly expanded and currently is the principal source of high quality aggregate and riprap in San Luis Obispo County. Large reserves of hard granite also exist elsewhere east of the Rinconada fault. All of the granite is more or less weathered to depths of 50 feet or more and must be stripped to expose the fresher rock. Other sources of hard, durable rock exist in smaller deposits within relatively uncrushed portions of the

Franciscan melange, the mafic volcanic rock unit, and diabase-gabbro of the ultramafic-mafic complex.

Other crushed stone materials used as imported fill or borrow, subbase, and surfacing of secondary roads have been obtained from weathered granitic rock, Franciscan melange, and the Toro, Atascadero, Monterey, and Santa Margarita Formations. Materials for such uses have relatively low specifications and can be obtained from most of the formations in the area.

REFERENCES

Addicott, W. O., and Galehouse, J., 1973, Pliocene marine fossils in the Paso Robles Formation: U.S. Geological Survey, Journal of Research, v. 1, n. 5, p. 509-514.

Atwater, T., 1970, Implications of plate tectonics for the Cenozoic evolution of western North America: Geol. Soc. Amer. Bull., v. 81, p. 3513-3536.

Bailey, E.H., Blake, M.C. Jr., Jones, D.L., 1970, On-land Mesozoic oceanic crust in California Coast Ranges: U.S. Geol. Survey, Prof. Paper 700-C, p. C70-C81.

Bailey, E.H., Irwin, W.P., and Jones, D.L., 1964, Franciscan and related rocks and their significance in the geology of western California: California Div. Mines and Geol., Bull. 183, 177 p.

Bandy, O.L., and Arnal, R.E., 1969, Middle Tertiary basin development, San Joaquin Valley, California: Geol. Soc. Amer. Bull., v. 80, p. 783-820.

Berkland, J.O., Raymond, L.A., Kramer, J.C., Moores, E.M., and O'Day, M., 1973, What is Franciscan?: Amer. Assoc. Petroleum Geologists Bull., v. 56, p. 2295-2302.

Bishop, C.C., 1970, Upper Cretaceous stratigraphy on the west side of the northern San Joaquin Valley, Stanislaus and San Joaquin Counties, California: California Div. Mines and Geol., Spec. Rept. 104.

Blake, M.C., Jr., 1970 (abstract), Different facies in Franciscan rocks and their significance in the late Mesozoic history of western California: Geol. Soc. America, Abstracts with Programs, v. 2, n. 2, p. 73.

Bramlette, M.N., 1946, The Monterey Formation of California and origin of its siliceous rocks: U.S. Geol. Survey Prof. Paper 212, 57 p.

Bramlette, M.N., and Daviess, S.N., 1944, Geology and oil possibilities of the Salinas Valley, California: U.S. Geol. Survey Oil and Gas Inv. Prelim. map 24.

Brown, J.A., Jr., 1968, Thrust contact between Franciscan group and Great Valley sequence northeast of Santa Maria, California: Univ. of Southern California, Ph.D. thesis, 234 p.

Burch, S.H., and Durham, D.L., 1970, Complete Bouguer gravity and general geology of the Bradley, San Miguel, Adelaida, and Paso Robles quadrangles, California: U.S. Geol. Surv. Prof. Paper 646-B, 14 p.

Calif. Dept. Water Resources, 1958, San Luis Obispo County investigation: State Water Resources Board Bull. 18, v. 1, 288 p. and v. 2, 13 appendices.

Calif. Dept. Water Resources, 1964, Crustal strain and fault movement investigation—Faults and earthquake epicenters in California: Bull. 116-2, 96 p., 3 map sheets.

Chipping, D.H., 1972, Early Tertiary paleogeography of central California: AAPG Bull., v. 56, n. 3, p. 480-493.

Compton, R.R., 1966, Granitic and metamorphic rocks of the Salinian block, California Coast Ranges, in Geology of northern California: California Div. Mines and Geol., Bull. 190, p. 277-287.

Curtis, G.H., Evernden, J.F., and Lipson, J., 1958, Age determination of some granitic rocks in California by the potassium-argon method: California Div. Mines and Geol., Spec. Rept. 54, 16 p.

Dibblee, T.W., Jr., 1968, Regional geologic map of San Andreas fault in Temblor Range, Carrizo Plain, and vicinity, Santa Barbara, San Luis Obispo, Kern, and Kings Counties, California: U.S. Geol. Surv., Open-File Report.

Dibblee, T.W., Jr., 1972, The Rinconada fault in the southern Coast Ranges, California, and its significance (abstract): Program to 47th Ann. Meet., Pacific Section Amer. Assoc. Petr. Geol., Bakersfield, Calif., Mar. 9-10, 1972, p. 39.

Durham, D.L., 1963, Geology of the Reliz Canyon, Thompson Canyon, and San Lucas quadrangles, Monterey County, California: U.S. Geol. Survey Bull. 1141-Q, 41 p.

Durham, D.L., 1965, Evidence of large strike-slip displacement along a fault in the southern Salinas Valley: U.S. Geol. Survey Prof. Paper 525-D, p. D106-D111.

Durham, D.L., 1968a, Geology of the Tierra Redonda Mountain and Bradley quadrangles, Monterey and San Luis Obispo Counties, California: U.S. Geol. Survey, Bull. 1255, 60 p.

Durham, D.L., 1968b, Geologic map of the Adelaida quadrangle, San Luis Obispo County, California: U.S. Geol. Survey, Map GQ-768.

Eckel, E.B., Yates, R.G., and Granger, A.E., 1941, Quicksilver deposits in San Luis Obispo County and southwestern Monterey County, California: U.S. Geol. Surv. Bull. 922-R, p. 515-580.

Evernden, J.F., and Kistler, R.W., 1970, Chronology of emplacement of Mesozoic batholithic complexes in California and western Nevada: U.S. Geol. Survey Prof. Paper 623, 42 p.

Fairbanks, H.W., 1898, Geology of a portion of the southern Coast Ranges: Jour. Geology, v. 6, p. 551-576.

Fairbanks, H.W., 1904, Description of the San Luis quadrangle: U.S. Geol. Survey, Geol. Atlas, Folio 101, 14 p.

Franke, H.A., 1935, Mines and mineral resources of San Luis Obispo County: California Jour. Mines and Geol., v. 31, p. 402-461.

Galehouse, J.S., 1967, Provenance and paleocurrents of the Paso Robles Formation, California: Geol. Soc. America Bull., v. 78, p. 951-978.

Gay, T.E., Jr., 1957, Specialty sands in Mineral commodities of California: California Div. Mines, Bull. 176, p. 547-564.

Gilbert, W.G., 1971, Sur fault zone, Monterey County, California: Ph.D. dissertation, Stanford Univ., 80 p.

Gilbert, W.G., and Dickinson, W.R., 1970, Stratigraphic variations in sandstone petrology, Great Valley sequence, central California coast: Geol. Soc. Amer. Bull., v. 81, p. 949-954.

Goldman, H.B., 1968, Sand and gravel in California—southern California: California Div. Mines and Geol., Bull. 180C, 56 p.

Hall, C.A., Jr., and Corbato, C.E., 1967, Stratigraphy and structure of Mesozoic and Cenozoic rocks, Nipomo quadrangle, southern Coast Ranges, California: Geol. Soc. Amer. Bull., v. 78, p. 559-582.

Hall, C.A., and Surdam, R.A., 1967, Geology of the San Luis Obispo-Nipomo area, San Luis Obispo County, California: Geol. Soc. Amer., Cordilleran Section Guidebook, Mar. 25, 1967, 25 p.

Hamilton, W., 1969, Mesozoic California and the underflow of Pacific Mantle: Geol. Soc. Amer. Bull., v. 80, p. 2409-2430.

Hamlin, H., 1904, Water resources of the Salinas Valley, California: U.S. Geol. Survey Water-Supply Paper 89, 91 p.

Harland, W.B., Smith, A.C., and Wilcock, B., eds., 1964, The Phanerozoic time-scale—a symposium dedicated to Professor Arthur Holmes: Geol. Soc. London Quart. Jour., Supp., v. 120s, 458 p.

Hart, E.W., 1966, Mines and mineral resources of Monterey County, California: California Div. Mines and Geol., County Rept. 5, 142 p.

Hart, E.W., 1971, Upper Mesozoic rocks near Atascadero, Santa Lucia Range, California, with special reference to K-feldspar content of the sandstone: Univ. of Calif. at Berkeley, M.A. thesis, 70 p.

Hart, E.W., in press, K-feldspar in upper Mesozoic sandstone units near Atascadero, Santa Lucia Range, California: California Div. Mines and Geol., Special Report.

Hart, E.W., in press, Limestone, dolomite, and shell resources of the Coast Ranges province, California: California Div. Mines and Geol., Bull. 197.

Hill, M.L., Carlson, S.A., and Dibblee, T.W., Jr., 1958, Stratigraphy of Cuyama Valley-Caliente Range area, California: A.A.P.G. Bull., v. 42, p. 2973-3000.

Hsu, K.J., 1969, Preliminary report and geologic guide to Franciscan melanges of the Morro Bay-San Simeon area, California: California Div. Mines and Geol., Spec. Pub. 35, 46 p.

Jennings, C.W., 1958, San Luis Obispo sheet: California Div. Mines and Geol., Geologic Map of California, Olaf P. Jenkins edition.

Kehle, R.O., 1970, Analysis of gravity sliding and orogenic translation: Geol. Soc. Amer. Bull., v. 81, p. 1641-1664.

Kellogg, R., 1921, A new pinniped from the upper Pliocene of California: Jour. of Mammalogy, v. 2, no. 4, p. 212-226.

Laizure, C. McK., 1925, San Luis Obispo County: Mining in California, California State Min. Bur. Rept. XXI, p. 499-538.

Lanphere, M.A., 1971, Age of the Mesozoic oceanic crust in the California Coast Ranges: Geol. Soc. Amer. Bull., v. 82, p. 3209-3212.

Loel, W., and Corey, W.H., 1932, The Vaqueros Formation, lower Miocene of California, Part 1—paleontology: Univ. Calif. Publ. in Geol. Sci., v. 22, n. 3, p. 31-410.

Logan, C.A., 1919, San Luis Obispo County in Report XV of the State Mineralogist—Mines and Mineral Resources of California, p. 674-726.

Loney, R.A., 1970, Faulting in the Burro Mountain area, California Coast Ranges, and its relation to the Nacimiento fault: Geol. Soc. Amer. Bull., v. 81, p. 1249-1254.

Maddock, M.E., 1964, Geology of the Mt. Boardman quadrangle, Santa Clara and Stanislaus Counties, California: California Div. Mines and Geol., Map Sheet 3.

McClure, D.V., 1969, Late Cretaceous sedimentation, southern Santa Lucia Range, California: M.A. thesis, Univ. California at Santa Barbara, 91 p.

McKenzie, D.P., and Morgan, W.J., 1969, Evolution of triple junctions: Nature, v. 224, p. 125-133.

Newsom, J.F., 1903, Clastic dikes: Geol. Soc. Amer. Bull., v. 14, p. 227-268.

Page, B.M., 1970a, Sur-Nacimiento fault zone of California—Continental margin tectonics: Geol. Soc. Amer. Bull., v. 81, p. 667-690.

Page, B.M., 1970b, Time of completion of under-thrusting of Franciscan beneath Great Valley rocks west of Salinian block, California: Geol. Soc. America Bull., v. 81, p. 2825-2834.

Page, B.M., 1972, Oceanic crust and mantle fragment in subduction complex near San Luis Obispo, California: Geol. Soc. Amer. Bull. v. 83, p. 957-972.

Reed, R.D.,1933, Geology of California: Am. Assoc. Petroleum Geologists, Tulsa, Okla., 186 p.

Richards, G.L., 1933, Geology of the Santa Margarita Formation, San Luis Obispo County, California: Stanford Univ., Ph.D. thesis, 173 p.

Ross, D.C., 1972, Petrographic and chemical reconnaissance study of some granitic and gneissic rocks near the San Andreas fault from Bodega Head to Cajon Pass, California: U.S. Geol. Survey Prof. Paper 698, 92 p.

Smith, P.B., 1968, Paleoenvironment of phosphate-bearing Monterey Shale in Salinas Valley, California: Amer. Assoc. Petr. Geologists Bull., v. 52, n. 9, p. 1785-1791.

Smith, P.B., and Durham, D.L., 1968, Middle Miocene Foraminifera and stratigraphic relations in the Adelaida quadrangle, San Luis Obispo County, California: U.S. Geol. Survey Bull. 1271-A, 14 p.

Taliaferro, N.L., 1943, The Franciscan-Knoxville problem: Am. Assoc. Petroleum Geologists Bull., v. 27, no. 2, p. 109-219.

Taliaferro, N.L., 1944, Cretaceous and Paleocene of Santa Lucia Range, California: Am. Assoc. Petroleum Geologists Bull., v. 28, p. 449-521.

Thorup, R.R., 1943, Type locality of the Vaqueros Formation in Geologic formations and economic development of the oil and gas fields of California: California Div. Mines, Bull. 118, p. 462-466.

Trask, P.D., and others, 1950, Geologic description of the manganese deposits of California: California Div. Mines, Bull. 152, 378 p.

Vedder, J.G., and Brown, R.D., Jr., 1968, Structural and stratigraphic relations along the Nacimiento fault in the southern Santa Lucia Range and San Rafael Mountains, California: Stanford Univ. Publ. Geol. Sciences, v. 11, p. 242-259.

Weaver, C.E., and others, 1944, Correlation of the marine Cenozoic formations of western North America (Chart no. 11): Geol. Soc. America Bull., v. 55, no. 5, p. 569-598.

Wilmarth, G.M., 1938, Lexicon of geologic names of the United States (including Alaska): U.S. Geol. Survey, Bull. 896, 2396 p.

Woodring, W.P., and Bramlette, M.N., 1950, Geology and paleontology of the Santa Maria district, California: U.S. Geol. Survey Prof. Paper 222, 185 p.

Manuscript submitted September 1972; revised June 1973.

www.ingramcontent.com/pod-product-compliance
Lightning Source LLC
Chambersburg PA
CBHW081232020426
42331CB00012B/3139